Praise for

WAVES IN AN IMPOSSIBLE SEA

"Matt Strassler has been one of the deepest thinkers in fundamental physics and quantum field theory for the past three decades. It is a cause for celebration to see him combine his penetrating insights together with a brilliant flair for beautifully clear and simple nontechnical explanations, to produce a true masterpiece with this book. I have never seen its equal and don't expect I ever will."

—Nima Arkani-Hamed, Institute for Advanced Study, Breakthrough Prize Laureate

"It's not easy to convey the ideas of modern physics without any equations, but also without compromises, making sure every statement is precisely correct. Strassler does it better than anyone I've ever read. If you want to know what's really going on in the realms of relativity and particle physics, read this book."   —Sean Carroll, author of *The Biggest Ideas in the Universe*

"This extraordinary work, reminiscent of the genius of Feynman, will awaken your sense of wonder and unveil the enchantment that surrounds our physical world. From the moment I delved into this captivating masterpiece, I found myself spellbound. It is a mesmerizing odyssey that will forever change how you perceive the world."

—Stephon Alexander, author of *Fear of a Black Universe*

"There is a particular zing you get from good explanations, and Strassler knows how to deliver them. This book is a rare attempt by a noted particle physicist to convey the core concepts out of which the world is constructed in language that truly anyone can understand. Strassler says he was motivated to write the book by the many egregious explanations he had read about how the Higgs field generates the masses of elementary particles— and, indeed, his version delivers the zing I've long sought."

—Natalie Wolchover, senior editor, *Quanta Magazine*

"Strassler succeeds triumphantly in conveying the fascination of the physical reality that underpins our world of atoms and stars. His distinguished expertise, combined with an entertaining and lucid writing style, enable him to lure readers into a 'deeper dive' than most physicists attempt when addressing a general readership—and to do this without distortion. He conveys the essence of the deep structures that underpin our natural world in an engaging and accessible way. This book deserves wide readership."

—Professor Martin Rees, Astronomer Royal

WAVES
IN AN
IMPOSSIBLE
SEA

WAVES
IN AN
IMPOSSIBLE
SEA

HOW EVERYDAY LIFE EMERGES
FROM THE COSMIC OCEAN

MATT STRASSLER

BASIC BOOKS

New York

Basic Books
Hachette Book Group
1290 Avenue of the Americas, New York, NY 10104
www.basicbooks.com

Printed in the United States of America

First Edition: March 2024

Published by Basic Books, an imprint of Hachette Book Group, Inc. The Basic Books name and logo is a registered trademark of the Hachette Book Group.

The Hachette Speakers Bureau provides a wide range of authors for speaking events. To find out more, go to hachettespeakersbureau.com or email HachetteSpeakers@hbgusa.com.

Basic books may be purchased in bulk for business, educational, or promotional use. For more information, please contact your local bookseller or the Hachette Book Group Special Markets Department at special.markets@hbgusa.com.

The publisher is not responsible for websites (or their content) that are not owned by the publisher.

Illustrated by Cari Cesarotti

Figure 13: Atomic resolution STEM imaging of perovskite oxide La0.7Sr0.3MnO$_3$. By Magnunor (Own work) CC BY-SA 4.0 (https://creativecommons.org/licenses/by-sa/4.0) via Wikimedia Commons.

Figure 31: "Wind Map" by Martin Wattenberg and Fernanda Viégas (hint.fm/wind).

Library of Congress Cataloging-in-Publication Data

Names: Sigmund, Karl, 1945– author.
Title: The waltz of reason: the entanglement of mathematics and philosophy / Karl Sigmund.
Description: First edition. | New York : Basic Books, 2023. | Includes bibliographical references and index.
Identifiers: LCCN 2023015358 | ISBN 9781541602694 (hardcover) | ISBN 9781541602700 (ebook)
Subjects: LCSH: Mathematics—Philosophy. | Mathematics—History. | Mathematics and civilization.
Classification: LCC QA8.4 .S5473 2023 | DDC 510.1—dc23/eng/20231012
LC record available at https://lccn.loc.gov/2023015358

ISBNs: 9781541603295 (hardcover), 9781541603301 (ebook)

LSC-C

Printing 1, 2023

In memory of Ann Nelson and Joe Polchinski,
mentors, colleagues, friends,
gone too soon

Contents

1 Overture   1

MOTION   11

2 Relativity: The Greatest Illusion   15
3 Coasting: Easier Than It Appears   29

MASS   43

4 Armor Against the Universe   45
5 Enter Einstein: Rest Mass   63
6 Worlds Within Worlds: The Structure of Material   79
7 What Mass Is (and Isn't)   97
8 Energy, Mass, and Meaning   106
9 That Most Important of Prisons   115

WAVES   123

10 Resonance   125
11 The Waves of Knowing   140
12 What Ears Can't Hear and Eyes Can't See   151

FIELDS   165

13 Ordinary Fields   169
14 Elementary Fields: A First, Unsettling Look   182
15 Elementary Fields: A Second, Humble Look   212

QUANTUM   221

16   The Quantum and the Particle   225
17   The Mass of a Wavicle   234
18   Einstein's Haiku   251

HIGGS   255

19   A Field Like No Other   257
20   The Higgs Field in Action   262
21   Basic Unanswered Questions   279
22   Deeper Conceptual Questions   290
23   The Really Big Questions   295

COSMOS   309

24   Protons and Neutrons   311
25   The Wizardry of Quantum Fields   320
26   Coda: The Extraordinary in the Ordinary   327

Acknowledgments   *333*
Glossary   *335*
Notes   *339*
Index   *359*

1

Overture

Imagine yourself clinging, like a character in a spy thriller, to the roof of a bullet train hurtling along at 150 miles per hour. Your situation is extremely precarious. As you are dragged along, the air resists your passage, and a hurricane-force wind threatens to push you off the back of the train. Your hair flies around wildly as you hang on for dear life.

And yet, as you read the opening words of this book, you may well have forgotten that you are sailing across the cosmos at 150 miles *per second*.[1] That's over 500,000 miles per hour. You are carried along with the Earth and Sun, in their orbit of our galaxy's center, thousands of times faster than the imaginary train. Nevertheless, you feel no space resistance. There is no "space wind" blowing your hair about. You travel through empty space as though it's not even there.

This wouldn't be puzzling if space were the benign, boring nothingness that we once thought it was. But after Einstein suggested that gravity reflects the bending of space and time, we learned that empty space itself can warp, stretch, and ripple. It is hard to imagine that nothingness could do these things, which seem more characteristic of materials such as fabric or rubber. Yet if space behaves like rubber, why can we move through it as though it's not there?

We do have some clues as to how empty space is distinct from rubber, air, or water. For instance, the waves in ordinary materials can always be overtaken: a speedboat can travel faster than ocean waves, and a plane can travel supersonically, outpacing its own sound. But you cannot catch up to ripples of space, known as "gravitational waves."

This would seem to violate common sense. No matter how quickly these ripples traverse the cosmos, you might imagine that a spacecraft with a powerful rocket engine could relentlessly increase its speed and eventually pass them by. But it just can't be done. Empty space is sort of like an ocean, and yet, in the end, it's not. There's something almost illogical about it.

As unusual as gravitational waves are, they're not alone; experiments have confirmed that light waves can't be overtaken, either. Moreover, they travel at the same rate as gravitational waves. These commonalities are striking, and in stark contrast to ripples in ordinary materials. Not only can sound waves, ocean waves, and earthquake waves be outrun, they each proceed at their own clip, as do ripples along ropes and in rubber sheets. So the fact that waves of light and undulations of space share remarkable properties suggests that the two might be profoundly interrelated. Perhaps they are different facets of a single, underlying structure.

It doesn't end there. Our own bodies are also made from tiny waves, mainly electrons and quarks, which we refer to as "elementary particles." Unlike light waves, these basic building blocks of ordinary material need not move at a fixed speed. This flexibility, crucial in allowing them to form atoms, rocks, and humans, arises from the fact that they have a property called "mass." They obtain their mass from a strange space-suffusing entity, an enigmatic presence known as the "Higgs field."

Yet though their speeds can vary, they are capped. Their motion cannot exceed the speed of gravitational and light waves—about 186,000 miles (300,000 kilometers) per second, or 670 million miles per hour—which seems to serve as an unbreakable cosmic speed limit. Why is there a single speed restriction that applies to all these different objects? Perhaps electrons and quarks, too, are a part of the same structure that incorporates light, gravitational waves, and space.

If so, what might this mysterious structure or system be, and how might it work? Our knowledge is limited. But we have a name for it. We call it "universe."

I don't just mean *The Universe*, as written with a capital "U" and spoken into a microphone with lots of echo—the gigantic black spaces that we typically block out of our minds except on clear, dark nights. I'm referring to the universe writ small as well as large, as it plays out in daily living: in our own bodies, in our homes, and in everything we encounter during every moment of every ordinary day.

Here's another curious fact, perhaps another clue. Obviously, you and I can't move through solid rock; we'd face stiff rock resistance, far more severe and destructive than the air resistance that would endanger us atop a bullet train. Yet *seismic waves*, waves in the Earth's rock caused by earthquakes and volcanoes, don't have this problem. They can travel directly across our planet from one side to the other facing no resistance whatsoever.

How do they manage this little miracle? It's not so mysterious. To the rock, our bodies are alien; the rock resists our presence in its territory. But seismic waves are vibrations of the rock itself. They belong there.

So what does it mean that we move through empty space—through the universe—without space resistance? Our drawings and descriptions of basic physics subtly lead us to imagine ourselves as made from ingredients that exist *within* the universe. But perhaps that's not so. It seems as though we are made from waves *of* the universe.

I do not mean this in a spiritual or metaphorical sense, though there's no harm in those resonances if you are inclined to hear them. My meaning here is concrete, tangible, *real* in the scientific sense. I am suggesting that our very substance is the cosmos in action. From this perspective, we are not merely residents of the universe, living within it as we live within our houses and apartments. Nor do we swim through the universe as fish swim through the sea. We are aspects of the universe, as seismic waves are aspects of rock and as sound waves are aspects of the air.

A better understanding of how the cosmos works, then, is a path to a better understanding of ourselves. We can gain insight into our senses, our muscles, our brains, our conception of what we are. Our connections to the outside world and to each other—our ability to see, hear, touch, interpret, communicate—become clearer. Central to all of these are fundamental though counterintuitive principles of physics, conventionally thought to be accessible only to experts. But perhaps it's time for conventional thinking to change.

My intention in this book is to bring these elements of the cosmos, and our place in it, within reach of a nonexpert reader, one who may have no background in science. But I'll be honest: the trail I've laid out is not a light walk in the park, for the universe's secrets are subtle and require serious thought. To paraphrase a quotation often incorrectly attributed to Einstein, everything in this book has been made as simple as possible, but not simpler.[2] There's no math, only concepts. I've avoided jargon wherever possible. I haven't assumed that you remember any science from school other than a vague flashback to a near-forgotten chemistry class. Nevertheless, you may find it helpful, as I did when I was first learning science, to read certain sections more than once.

Why did I write this book, and why do I hope you'll read it? There are many answers, some of which I'll come to later. But here is perhaps the most important one. If you, like me, harbor deep and existential questions regarding why we are here and what life is about or ought to be about, and if you stare into the empty eyes of the night wondering what it means to be a human being, then I suspect you might find insight, more than you may imagine, through a better understanding of how the universe functions within us. A personal lesson that I myself have learned, drawing upon my long experience as a physicist, is this: it is only with a clear image of how mind and body intersect with the world that one may hope to find a road to thorough self-knowledge, and to a full appreciation of what it means to be alive.

Although this book's purpose is to illuminate how the most esoteric-seeming physics affects every aspect of human existence, my initial goals were less ambitious. I was originally motivated by a simple fairy tale, a seemingly harmless little lie.

In 2012, physicists at the world's largest particle accelerator, the Large Hadron Collider (or LHC for short), discovered a long-sought type of particle called the *Higgs boson*. The media enjoys calling it the "God Particle."

But most particle physicists, including Peter Higgs himself, think this name is a bit silly. Higgs bosons play no role in daily life or in the wider cosmos. You won't find them lying on the ground or wandering between the stars, and they haven't done anything of interest since the early moments of the universe. The reason is simple: a Higgs boson, once created, disintegrates in a billionth of a trillionth of a second.

This is why physicists needed the LHC in the first place. In order to have any hope of discovering these elusive beasts, we humans had to try to make new ones from scratch. But why bother to make these ephemeral particles at all? This was an important question, since building the LHC and its predecessors took a great deal of money and time.

The answer is that the search for Higgs bosons wasn't an end in itself. Instead, it was a means to a far more important end. The rationale for the endeavor was that finding the Higgs boson would prove the existence of something of much greater significance: the Higgs *field*.

This field, unlike the corresponding particle, is long-lasting and has been a cosmic presence since the universe was born. Over billions of years, it's been switched on, steady, constant, and uniform across the entire visible universe—around Earth, within Earth, and within us, too.

While it is sometimes said that "the Higgs field gives everything in the universe its mass," this is a considerable overstatement. But still, the Higgs field is responsible for the masses of certain crucial elementary particles, including the electrons found in every atom. If electrons had no mass, atoms would never have formed, and neither we nor the Earth would exist. Thus, the importance of the Higgs field is beyond debate. Our lives depend upon it.

Learning this, curious journalists and politicians asked the physicists further questions. "How does it work, this Higgs field? How does it give things their mass?"

By the time you reach the final third of this book, you'll know the answers. But the journalists and politicians weren't asking for a book; they wanted a quick reply, a sound bite. To satisfy them, a little story was invented.

I'm hesitant to call this story an outright lie; its inventors and purveyors were well-meaning and weren't seeking to mislead anyone, even though they knew what they were saying wasn't really true. I can't really call it a myth or a fable or a fairy tale, either. It's a very special type of falsehood common in explanations of physics for nonexperts, so I'll call it a physics fib or, more simply, a phib.

Phibs are often found in articles and books about the universe. They arise when well-intentioned physicists, faced with a nonexpert's question, are trying to concoct a short, memorable tale to serve as a compromise between giving no answer at all and giving a correct but incomprehensible one. This is a challenge that physicists often confront, especially when meeting with politicians or journalists who want at most a paragraph and perhaps no more than a sentence. Typical phibs are mostly harmless and are quickly forgotten. But sometimes a phib spreads widely and is taken far more seriously than its author ever intended. Then it may do more harm than good.

The Higgs phib has a number of variants. Here's a short version of one of them: *There's this substance, like a soup, that fills the universe; that's the Higgs field. As objects move through it, the soup slows them down, and that's how they get mass.*

It's remarkable that such a short story can be wrong about so many things at once—wrong about the "soup," wrong about mass, wrong about motion. As we'll see later, it involves a sleight of hand that makes it sound far more sensible than it is. But should it bother us when particle physicists misrepresent this detail of their research? I'll try to convince you that it should.

For one thing, as I've just explained, the Higgs field is more than a detail; it belongs on the top-ten list of essential ingredients for life. Something so foundational to existence ought to be explained properly, it seems to me.

Yet there's an even more important issue. This apparently innocent yarn about slowing and soup tears a hole in the heart of a cosmic principle, one that lies at the core of our conception of the universe.

At stake is nothing less than *the principle of relativity.*

This principle is arguably the most durable of all known laws of physics. It has had broad historical and cultural significance, too. Occasionally suggested over the millennia, only to disappear repeatedly in a cloud of confusion, it was finally put on a firm footing by the icons of modern physics:

Galileo Galilei (1564–1642), Isaac Newton (1642–1727), and Albert Einstein (1879–1955). Without it, the universe is rendered incomprehensible.

Simply put, the Higgs phib butchers the relativity principle. This makes its explanation of the Higgs field's role in nature—or rather, its pretense of an explanation—completely counterproductive, in that it diminishes human understanding rather than augmenting it. We are led to wonder what mission it actually serves.

To be sure, describing the Higgs field requires more than a sound bite. It requires a book, the one I originally intended to write. But to explain how the Higgs field does its job, I had to draw together many of the most important concepts of modern physics, from Einstein's time to the present day. And so, as this book took shape, I found its aspirations becoming more lofty, extending beyond its initial aims to encompass physicists' contemporary view of the cosmos.

In an effort to convey that worldview to a broad audience, I have tried to make this text largely self-contained and nontechnical. (Inevitably, there are topics and technicalities that can't fit within its pages; asterisks in the endnotes indicate subjects that I have explored further on my website, whose address is given at the back of the book, prior to the endnotes.) By the end, we will encounter some of the most startling and sophisticated issues in modern physics, ranging from the nature of space and the role of the Higgs field to the existence of atoms and of macroscopic objects made from them. But we will start with ideas that long predate Einstein.

The book's first third will explore a few foundational concepts from a modern perspective. These notions—motion, mass, and energy—pervade our daily lives. For those who've read about physics or even studied it, this may seem familiar territory, but I'll draw your attention to critical details that are often overlooked or scrambled. Central to this opening section will be Galileo's version of the principle of relativity. Then we'll jump three centuries to Einstein, his updated version of relativity, and his most famous (and often misunderstood) formula. As we come to see how mass, motion, and energy are intertwined, we'll encounter challenging puzzles concerning the origin and nature of mass, especially that of the electron.

It may not be immediately obvious how these puzzles relate to the book's middle third, which begins with vibrations, waves, and the fundamentals of music. After a brief survey of the physics and physiology of sound and light, we will turn to the waves of the universe itself. This will bring us to the subtle subject of fields. Even for those who have learned about fields in a first-year physics class, this material will cover new ground, because the perspective I will take differs from that of most physics courses. Though we won't fully explore Einstein's view of relativity, I'll draw your attention to the strange nature of space and time and to the importance of Galileo's relativity principle in Einstein's thinking.

Physicists' understanding of fields is both profound in some senses and quite limited in others. Because of this, I will have to leave certain obvious and important questions unanswered. I hope to make clear both what we know and what we don't.

The last third of the book enters the quantum realm. We will not need to dive deeply into the most confusing intricacies of quantum physics; instead, we will focus on key principles. After clarifying the relation between particles, waves, and fields, we will solve a variety of mysteries from earlier in the book. The nature of the electron's mass and that of other particles will finally be revealed, along with the reason that all electrons are literally identical. Then, assembling insights from many previous chapters, we will learn what it really means to say "the Higgs field gives the electron its mass."

The discovery of the Higgs boson confirmed the existence of the Higgs field, resolving some long-standing questions about the universe, but it left other puzzles unaddressed and posed many new ones. After describing and exploring these unsolved problems, I'll conclude the book by considering how the cosmos and quantum physics intersect with one another and with the everyday world. By highlighting the ways in which these peculiar features of the universe affect our lives, I hope to offer you a clearer sense of how we fit into the cosmos and of how the ordinary emerges from the inconceivable.

The worldview I'll describe here is one I came to over decades as a theoretical physicist. It has been shaped by years of physics research, naturally, but other factors have also played a role. Growing up in a rural part of the United States, in the state of Massachusetts, where I live and work today, I had a childhood of star-filled dark skies, towering trees, and animals both wild and domestic. Those early experiences influenced how I view nature and the place of humans within it. Another constant in my life has been a love of music, which plays a central role in this book.

During my career as a professional physicist I worked at several universities and research institutes, investigating the nature and behavior of particles, fields, and strings. I taught in settings formal and informal, explaining physics to undergraduate students, adult nonscientists, budding experts, and personal friends. At a certain point I went into semiretirement, and, while continuing to do research and train young physicists, I turned to blogging and other ways of communicating science to a broad audience. This was a natural step for me; I've always loved telling people about the amazing universe we live in, whether they've wanted to hear about it or not. (My first science lecture, about the planets, was given voluntarily at the age of five. "It's very cold on Pluto," I told my fellow kindergarteners.)

Finally, I have turned to the writing of a book. But my choice of subject might seem surprising. It's been more than ten years since the Higgs boson's discovery and over a century since Einstein's great breakthroughs. You might well wonder whether there's anything left to explain.

I think there is. What's been missing, to my mind, is the full story of how modern physics and human life fit together. It's not easy to tell that story. To do so without relying on phibs requires breaking down and repackaging concepts that at first appear technical and impenetrable. But it seems I have a knack for repackaging the impenetrable—luckily for me, for without that skill, I could not have been a successful physicist. I have always been surrounded by people much smarter than I am. If I hadn't quickly found ways to disentangle their complex ideas, I would never have mastered the subject.

I hope this skill has borne fruit in this book, in which I have tried to create an account that is both comprehensible and accurate, doing my best to avoid exaggeration, speculation, and phibs. In presenting a contemporary viewpoint common among professional physicists, I've tried to satisfy the desire of many readers, a desire I know is out there, for a straightforward, honest depiction of the cosmos as best we understand it. By shining as clear a light as possible on what we know, I hope I'm also demarcating the darkness—the edges of that yawning abyss of ignorance that draws physicists of the present ever onward, and into which physicists of the future, including perhaps some of you, will carry a lantern.

MOTION

To be is to move. We are never stationary for long; living requires us to seek food, resources, companionship. Even when still, we continue to breathe; our hearts beat, our blood flows, and electrical currents run through our nervous systems. In every cell, the reading of our DNA and the carrying out of its instructions require motion at the molecular scale. And when we look down into the subatomic realm, we find that every fragment of our bodies is forever spinning, roving, vibrating.

We take motion for granted. Were it not for the insights of physics, we might never have noticed that it hides mysteries as deep as any that we confront in life.

Secrets and illusions permeate the human experience of the world, and our struggle to overcome them forms an important chapter in our species' history. Foremost among cosmic secrets is the roundness of the Earth. Over two thousand years ago, Greek thinkers became experts in geometry and found clever tricks for estimating the Earth's shape and size.[1] Their discoveries were soon widely known not only in ancient Greece and Rome but also in India, across the Islamic world, and elsewhere. Despite what some schoolbooks still claim, educated Europeans were well aware that the Earth is round, even before Columbus, Magellan, and other explorers of the Renaissance set sail.

In our era of satellites, intercontinental shipping, and air travel, not to mention photographs from outer space, it is amazing that anyone could doubt that our planet is ball-shaped. Numerous technologies, including the Global Positioning System (GPS), widely used for navigation, rely upon it. Admittedly, though, Earth's shape is not *intuitively* obvious, and that's the problem. Despite occasional hills and valleys, the ground around us appears to stretch out like an approximately flat surface, as does the ocean on a calm day. It would be easy to take the Earth's apparent flatness for granted were we not taught otherwise as children.

A simple fact of geometry explains this illusion. Any huge sphere will seem flat to tiny creatures that roam its surface. These creatures must transcend the limitations of their senses, using thoughtful observation and logical reasoning, if they are to recognize their intuition as naive.

As a species, we hold tightly to our intuitions, and we tend to believe them. But recent centuries of science have taught us that most assumptions we typically make about the material universe are founded upon misconceptions. This is among the most important lessons of human history: we must never ignore facts, but when we try to interpret them we must beware, for common sense is a thoroughly unreliable guide to the workings of the natural world. No matter how strong an intuition may be, we must be prepared to let it go.

Take for example the sensation of lying quietly in bed or sitting relaxed in a chair, as one might do while reading a book. It's peaceful and still, perfect for a little nap.

And yet that sense of resting peacefully is a mirage. You and I and the Earth are careening along at over 150 miles every second as our planet and the Sun orbit the center of the Milky Way galaxy, the city of nearly a trillion stars that is our cosmic home. Each minute, the Earth travels (relative to the galaxy's center) a distance comparable to its diameter (Fig. 1). Every second, we are carried across a span that by car would require a couple of hours, as from Philadelphia to New York, from Zurich to Basel, from Beijing to Tangshan. At this rate, more than twenty times the rapid pace of artificial satellites orbiting the Earth, we could circle our planet in under

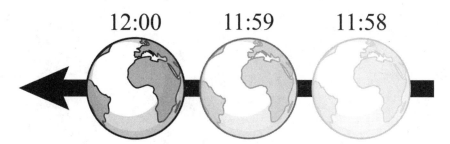

Figure 1: As seen from our galaxy's center, the Earth (shown at three locations one minute apart) travels at tremendous speed.

three minutes, land on the Moon in half an hour, and reach the Sun in a week.[2]

For tens of thousands of years, humans hadn't the faintest idea that we roam the cosmos. Even once we suspected it, we could not easily guess our speed and direction. Only in recent decades have our motions, relative to our own galaxy and to other galaxies, become clear.

2

Relativity

The Greatest Illusion

The fact that we aren't aware of our spectacular velocity reflects another great cosmic secret. We can't sense this motion because steady motion in a straight line—travel at a constant speed and in a constant direction—cannot be detected in our universe.

Specifically, suppose you are in a closed room and can't look outside. Then it is impossible, by pure feel or by any scientific measurement, to distinguish being in smooth and steady motion from being perfectly stationary. Nor can you figure out how fast you might be moving or in what direction. It cannot be done. Period.

This is the *principle of relativity*, or at least its most elemental, durable, and disquieting part. To put it another way, *in quiet, undisturbed conditions, within an isolated bubble with no access to the outside world, there is absolutely nothing you can do to establish either the amount or direction of your motion.*

As stated, this principle might seem abstract to the point of irrelevance. The ideal isolated bubble would be a thick-walled, windowless spaceship far out in interstellar space, gliding gently with its rocket engine switched off.[1] Such a craft is fun to think about, perhaps, but few if any of us will ever travel in one. We might well question the merit of putting something so remote from human experience at the core of science.

Yet nearly isolated bubbles play a surprisingly large role in our lives. An example is the Earth itself. It's not completely isolated, and careful scientific

15

experiments can measure Earth's spin and its motion relative to nearby planets and distant stars. But those experiments are challenging; neither you nor I can perform them with our senses, or even with simple equipment such as portable amateur telescopes. And so for us, in daily life, the Earth does act as though it were an isolated bubble. That's why our incessant and rapid motion goes unnoticed.

Other nearly isolated bubbles include a more realistic spaceship with thin walls and windows, or even an airplane in tranquil air, especially if we're sitting far from a window or looking out into the night over a dark ocean or cloud deck. The principle of relativity explains why simple experiences of life—breathing, walking, drinking—are unaltered inside such a plane. Even a train or car can serve as a bubble if its motion is straight and smooth, the windows are closed, and your eyes are shut. It's true that if you take advantage of all the clues around you, you usually don't need a fancy experiment to tell you that your plane, train, or car is moving relative to the ground. But the relativity principle assures that when you restrict your actions to the interior of a smoothly coasting vehicle, and you fail by choice or accident to take in information from outside it, then your informal experience inside that conveyance will be just as though it were an isolated bubble.

Meanwhile, the relativity principle has a surprising amount of influence at the atomic and subatomic levels. Atoms and other collections of subatomic particles often act (briefly) as though they are isolated. That's why relativity is important not only for astronomers but also for particle physicists.

So yes, we do often encounter isolated bubbles, albeit approximate and temporary ones. To the extent that we're within one, we can observe some of the consequences of the principle of relativity. But even then it takes a concerted effort. That's because we are never isolated from other objects that are with us inside the bubble: floors, walls, chairs, tables, air, water. Our intuition about the world comes from our interactions with these types of objects, which are remarkably effective at obscuring the relativity principle and distracting us from its implications. They conceal precisely those aspects of the cosmos that would otherwise help us make intuitive sense of it.

Although the relativity principle is easily stated in a few words—steady motion is undetectable—it runs counter to human psychology. It violates assumptions about the world that all of us, including future physicists, develop as children. It's almost as though daily life were designed to put basic physics out of the reach of the human mind.

This is why the relativity principle escaped even the brilliant mathematicians and philosophers of ancient Greece. Though they proved that the Earth is a sphere and measured its size without traveling far from home, they never concluded that the Earth moves. A few individuals did suggest that the Earth spins and travels, but the most influential thinkers, believing any such motion ought to be easily perceived, argued otherwise. It took many more centuries for humans to learn that motion need not be easily felt. Our planet rotates and roams the heavens, but our motion is *nearly* steady. That makes it *nearly* undetectable, thanks to Galileo's principle.[2]

Really, it's an underappreciated triumph that our species ever managed to overcome this psychological obstacle. To do so required a series of our greatest thinkers, building on each other's insights.

2.1 Galileo's Ship

Today, the concept of relativity is commonly associated with Einstein's notions of space and time, developed in the first decades of the twentieth century. But the issue of relativity goes back centuries before him. It addresses fundamental questions about reality: Does a particular way of looking at the world, or a certain property of an object in the world, depend on your perspective? If it does, how? If it does not, why? Or, to put it more scientifically, which aspects of the universe are *relative*—dependent on an observer's perspective—and which ones are not? And for those that are relative, how exactly can you translate between one person's perspective and another's? These were already questions for Galileo, and he gave them initial answers long before Einstein came along and amended them.

Galileo articulated the principle of relativity after performing a series of experiments on motion. In his book *Dialogue Concerning the Two Chief World Systems*, published in 1632, he explained it to his contemporaries in literary form. "Shut yourself up with some friend in the main cabin belowdecks on some large ship," it begins, and the ensuing five hundred words may be boiled down to a single sentence: *a person belowdecks on a smoothly sailing ship cannot hope to determine whether the ship is in motion or, if so, what is its speed.*

Beautifully expressed as Galileo's argument is, his seafaring isolated bubble must have seemed abstract to the point of uselessness to most of his contemporaries, almost as abstract as my spacefaring bubble seems to us. Most people experienced travel only by foot, horse, or cart, which are neither smooth nor protected from the air flowing by. To appreciate Galileo's insights and the true nature of the world, they would have needed to imagine what boat travel is like belowdecks. Even today reality is so obscured by the complexity of ordinary life that we need imagination to recognize how simple it truly is. In this, there is considerable irony.

Progress after Galileo's insight was hardly instantaneous. Decades passed before Newton, embracing specific ideas of René Descartes, Christiaan Huygens, and others, built his comprehensive understanding of motion upon Galileo's principle. This foundation of physics remained stable for two centuries, until Einstein realized that even the time that elapses between one event and another is a matter of perspective. Yet despite Einstein's revolutionary ideas, he preserved the central principle that steady motion is undetectable. This core precept of relativity may well be the oldest law of physics never to have been rejected or significantly improved.

Despite popular lore, Einstein certainly did not say that "everything is relative." Such a statement is false, in fact. As we'll see in this book, there are a number of concepts that aren't relative—everyone agrees on them—and they are among the most reliable aspects of the universe.

But one thing that's certainly relative is *speed*. In steady motion, no one can justifiably claim that "I'm moving and you're not" or "you're moving and I'm not" or "we're both moving." Such statements are mere matters of perspective.

If you're sitting on a park bench, you may see yourself as stationary. If I drive past you, traveling north at 40 miles an hour, my perspective is that I'm stationary in my car, while you and your bench are moving south at 40 miles an hour. From the perspective of someone sitting on the Moon, we're both moving at hundreds of miles per hour as our planet spins. That's the thing about different perspectives: no one's point of view is in any sense better. They're just different ways of viewing exactly the same thing. When something's relative, everyone disagrees, yet no one is wrong.

Galileo's principle takes advantage of this relativity of motion. Inside a steadily moving isolated bubble, where you have no view of, contact with, or perspective on the outside world, your motion is undetectable *because it has no perspective-independent meaning in our universe.*

This is not in any sense obvious. Certainly it wasn't to me before I read books on the subject. It's not an accident that it took the greatest of minds to figure out how relativity works.

Here's another strange aspect of motion. Each of us, lying in bed or sitting at our desks, may feel that we're stationary. But in fact, we're all moving relative to one another. That's due to the Earth's rotation, which carries us along at different speeds and directions; see Fig. 2. As seen from Earth's center, those of us near the equator travel faster than those of us near the poles; those of us on opposite sides of the Earth move in opposite directions. More generally, two people at the same longitude but different latitudes travel at different speeds around the Earth, while two people at the same latitude but different longitudes travel around the Earth at the same speed but in different directions.

Unlike motion in a straight line, to which Galileo's principle applies, motion in a circle is often easy to detect. Remember, back in childhood, when an adult swung you around and around? You could certainly feel it even if your eyes were closed. The same goes for circular rides in amusement parks. Each of us makes a daily circle around the Earth's axis, so why can't we perceive that motion?

The reason is one I just mentioned: because the relativity principle assures that straight-ahead steady motion is completely undetectable, it assures that

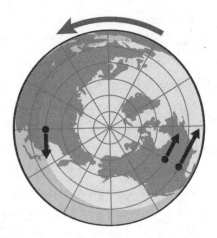

Figure 2: Sleeping babies are all in motion relative to one another. As
seen from our planet's north pole, their speeds and directions (black arrows)
vary with their latitude and longitude. Any one location on the planet
will be seen, from another location, as traveling in a daily circle.

steady motion in a nearly straight line can't easily be felt, either. Our daily
paths around the Earth curve very gently, differing from a straight line by
just one degree every four minutes. That's far too gradual for us to notice.
Just as a huge planet seems flat to tiny creatures on its surface, steady motion
on a giant circle around a slowly rotating planet feels much like steady motion in a straight line.[3]

This helps explain why we each feel stationary when we're sitting or lying
down. We are oblivious to our own motion, and also to the relative motion
between ourselves and our friends in other parts of the world. That relative motion isn't slow. If people sitting in Boston were to measure carefully,
they'd see people standing in Miami as moving at 215 miles per hour; meanwhile, those in Miami would perceive their friends in Boston as moving at
215 miles per hour in the opposite direction.

But wait: The distance from Miami to Boston, 1,257 miles, never changes,
so how can there be relative motion between those two cities? It's because
Bostonians view Miami as moving in a daily circle, one that leaves the distance between the two cities always unchanged—and vice versa. You can get
a hint of this from Fig. 2; if you turn the picture in a circle centered on any

one of the black dots, you'll see that dot as stationary while the other two dots move around it.

The same Bostonians are viewed as moving at 689 miles per hour by people working in San Francisco, 825 miles per hour by people at a London pub, and 1,517 miles per hour by people half asleep in Sydney. In each case, the reverse is true, too; Boston folks regard their distant friends as the ones moving.

Because these speeds, like all motions, are relative, the wide diversity of opinions poses no contradiction. Everyone is right. The Bostonians, who think themselves stationary, are seen quite differently by people scattered around the globe. The same is true for each of us no matter where we are. We're all in motion relative to our distant friends and family, even when we think we're going nowhere.

A friend of mine, whom I'd met for coffee, tried to wrap her head around this. "So you're saying that everyone around the world who is sitting down and thinks they're stationary—they're all wrong?"

"It's not that they're simply mistaken," I explained. "To say 'I am stationary' is meaningless."

She gave me a confused look.

"It's the same as with any other word that's a relative term," I pointed out. "It's like someone describing me as tall."

"Umm..." she tittered. "I wouldn't have said that...."

"But relative to whom?" I asked. "I'm tiny relative to redwood trees, but to a mouse I'm a giant. That makes me both tall and short at the same time. And that, in turn, makes it impossible to claim unambiguously that I'm either one.

"Sure, when someone says, 'I'm tall,' they usually mean, implicitly, 'I'm tall relative to the average human.' And when someone says, 'I'm stationary,' they implicitly mean 'I'm stationary relative to the objects in my immediate surroundings.' But without a context, statements like 'I'm tall' or 'I'm strong' or 'I'm loud' have no meaning. In the same way, simply saying 'I'm moving quickly' or 'I'm not moving at all' has no meaning in a universe like ours, in which all speed is relative and steady motion can't be detected.

One can imagine universes in which such statements might make sense. But they're meaningless in this one. Your motion always has to be expressed as relative to some other person or thing."

She pondered this for a few moments. "So when I say my car's going at 60 miles an hour, you're saying I'm secretly comparing the car to the road it's on. And not to roads on another continent, compared to which it would be moving at some other speed. Is that your point?"

"It's part of the point," I said. "Another part is that the car isn't moving at all relative to its driver and passengers. So it's stationary and moving, and it's fast and it's slow—just as I'm tall and I'm short."

"But then, is there *anything* in the universe that's truly stationary?" she ventured.

"It's impossible for an object to be stationary, and it's impossible for it not to be stationary," I insisted. "You are always stationary with respect to yourself and generally some other objects around you, such as your shirt, but you are always moving relative to most things in the universe and even to most things on this planet. And you are moving at many different speeds and directions relative to those things."

Our coffee drinks appeared at the bar, so we paused briefly to retrieve them. I'll take this moment to admit that what I'd just said to my friend— that we're always both stationary and not stationary—is not exactly true. It would be 100 percent true if we were in steady motion in straight lines, but when we're moving in a tight circle, as when we round a sharp curve, we can tell we're not stationary. Nevertheless, when sitting or moving steadily upon an immense, slowly rotating planet, in circular motion that's so nearly straight over minutes that we can't sense it, my remark is essentially true, both for practical purposes and as far as it affects our daily experiences. And it's 100 percent true that our motion is always ambiguous; we cannot ever say what it is without stating it relative to something else.

"You know," I continued as we sat down again, "it's really hard to express these ideas clearly. I mean, all this about being stationary and moving at the same time, and not being able to say which direction you're going in or how fast…if you didn't know better, you might think I was crazy. It's almost impossible to describe it using sentences that sound logical, partly because we just don't have the right words and concepts in our language."

"Well," she countered, "that's not very surprising, is it? We rarely have words for things that we don't actually experience."

"What do you mean?" I exclaimed, spreading my arms wide. "We never experience anything else!"

She stopped short, her expression frozen. Then, after a long moment, she started to laugh.

"Gosh, this is hard to keep straight. But I'm getting there, and I think I'm starting to see your point. And maybe you do need a word for it. What about..." She paused. "What about *polymotional?*"

"Hmm," I replied. "That's not bad! Or maybe even *omnimotional*. Pick any speed and direction you like; that's our motion relative to some particle out there in the universe." There are hordes of subatomic particles flying about the cosmos. Choose any one of them. From our perspective, it's moving and we're not, but who is to say it's not the other way around?

"*Ambimotional?*" she offered.

The conversation brought to mind a famous line composed by Canadian humorist and economist Stephen Leacock:

> Ronald flung himself from the room, flung himself upon his horse, and rode madly off in all directions.[4]

2.2 Relativity and Intuition

All sorts of common experiences make the relativity principle counterintuitive. Under normal circumstances, we can usually tell when our car or train is moving across the ground; we sense the vibration and noise that come from a rubber tire rolling on asphalt or a metal wheel moving on an imperfect rail. But this noise and vibration aren't caused by the motion itself. They arise from the wheels rolling on the asphalt or on the rail—from direct contact between a part of the vehicle that's moving in one way and something on the ground that's moving in a different way. Suppose that this contact between vehicle and ground were somehow removed, as in a magnetically levitated train. Then it would become extremely difficult to guess, with eyes closed, whether we were in motion, how fast we were going, or in which direction.

Try for a moment to imagine what it would be like if we were in outer space on a spaceship. Then there'd be no wheels to cause noise and no road to cause vibration. The motion across the emptiness of space would be smooth and silent, and there'd be no clues as to our steady motion.

In fact, no imagination is needed. Just look around you. We are already in outer space on a spaceship, which we call Earth. Its swift motion produces neither noise nor vibration, which is why we don't notice it.

When I pointed this out to a friend, he expressed disbelief. "Spaceship? But the Earth doesn't have a rocket!" The analogy between an artificial spaceship and the Earth rang false to him. But my friend had been misled by confusing and confused movies that imply that spaceships fire their rockets to keep themselves moving. This isn't true.

The rockets on a spaceship are needed only to speed up, slow down, or change direction.[5] Once the craft is moving as desired, the rockets are extinguished and are unneeded for most of the trip. With its engines off, the ship cruises through space, moving without vibration or other disturbance.[6]

This contrasts with airplanes, which always battle air resistance, and with cars, which battle friction from the road, from air turbulence, and from their internal moving parts sliding past one another. An airplane without a running engine must glide to the ground to avoid crashing; a car with its engine turned off will soon grind to a halt. Not so a spaceship.

Often science fiction films and television shows get this wrong. For example, there's an episode of *Doctor Who* (mild spoiler alert) in which the Doctor, visiting a large spaceship on a long voyage, notices that there's no vibration throughout the ship. From this he deduces that the spacecraft has no running engine. As this seems impossible to him, he concludes that it must be traveling through space via some unconventional means.

Well, this made me chuckle, because the poor Doctor gets the principle of relativity exactly backward. In fact, the presence of vibration, rather than its absence, would have been a clue that something was amiss. No engine is necessary merely to cruise at a steady speed through empty space; just ask the Earth.

The writers of the episode applied common sense, obtained from our experiences of motion through air and water, to motion through empty space, where such intuition goes badly awry. But my point is not to criticize them. Their errors are so natural! The seventeenth-century genius Johannes Kepler held similar misconceptions. Besides, *Doctor Who* is science *fiction*; its very premise involves scientific inconsistencies. A few misunderstandings of the cosmos are worth a good story. Nevertheless, this story and others like it reinforce the psychological assumptions that make science *fact* so difficult for humans to grasp.

As for the Earth, it never needed a rocket or any other means of propulsion. It was born in motion, emerging along with our solar system's other planets out of a spinning disk of dust that surrounded the infant Sun. Its engineless travels will continue, influenced only by gravity's weak pull, until a bloated, dying Sun brings them to an end.[7]

Airplanes, unlike spaceships, run their engines continuously. They need the air to keep them afloat, but at a price: they must always push their way through it, and so they're forever fighting air resistance.

Still, inside the plane you're protected from that air resistance, and the aircraft's interior acts as though it were an isolated bubble. You can tell the plane is moving if the air is turbulent, making the motion unsteady and unpredictable, and you can feel when the plane is speeding up, slowing down, or beginning to climb or descend. But when its motion is steady enough, you will not be able to prove that you're moving at all.

A reader of my blog related the following anecdote. "I had the experience," he wrote, "of waking up after a nap on a big jet—one of those monster A380s that's incredibly quiet if you're on the upper deck—and it was so peaceful that it took me half a minute to realize that we were still in flight. I thought I'd slept through the landing!"

Even on a louder plane, you can have fun trying to guess how fast you're going. If the plane's motion is steady, there's no way to tell (without looking outside—no cheating!—or listening very closely to the air rushing by the fuselage) whether your airspeed is 200, 300, or 400 miles an hour. Life feels perfectly normal inside the plane; at any constant speed, you can play catch in an airplane aisle as easily as on the ground.

Here's another game: go to the back of a plane, close your eyes, and turn around a few times while trying to forget your original orientation. Then, before you open your eyes, try to guess the direction in which the plane is headed. It's not easy, thanks to the principle of relativity.

Or if you're sure nobody's watching, jump straight up as high as you can. You'll find that your jump feels the same as when you're at home; you'll come straight down again, relative to the plane's floor. Don't ask me how I know this. (Okay, okay; I was eleven. I was quite surprised at the outcome.) Even though the plane is moving toward its nose while you're in midjump, the rear wall of the plane will not approach you, any more than a wall of your house would approach you if you jumped straight up in your own bedroom. In both cases, your experience will be exactly as shown at the left of Fig. 3. Meanwhile, a person on the ground watching you would see your jump as forming an arc, as shown at right in Fig. 3, but the plane would move with you in just such a way that at all times you'd remain above the same spot on the plane's floor.

Grasping the relativity of speed is crucial for pilots, who must separately keep track of their *ground speed* and their *airspeed*. The airspeed, the speed of the wings relative to the air, determines whether the plane can fly; if that speed's too low, the plane will fall out of the sky, and if it's too high, the plane will break apart. Airspeed determines when a plane can take off because it's the air rushing over the wings that provides the lift that allows the plane to rise. But it's ground speed that determines how quickly a plane is approaching the end of the runway and how quickly it moves from its departure airport to its destination. The two speeds can differ significantly in strong winds. Once, on my way from the New York area to visit the LHC in Geneva, I flew across the Atlantic Ocean at nearly the speed of sound! Had that been our airspeed, my plane would have disintegrated. But the flight was perfectly safe; with the aircraft pushed along by a tailwind roaring at 200 miles per hour, only its ground speed was unusual.

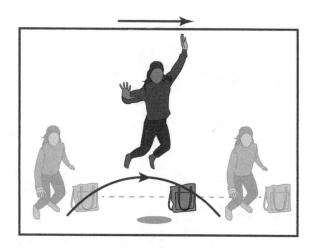

Figure 3: If you're at home and jump straight up (left), you will land where you started; the bag sitting a foot away will still be there when you land. If you jump in an airplane overhead, your own experience of the jump will be the same as at home (left). Someone on the ground will see your jump as an arc (right), but your motion will seem synchronized with the plane's motion, assuring that you jump and land at the same spot on the plane's floor.

Relativity also explains why airplanes take off into the wind whenever possible, as illustrated in Fig. 4. For the wings to generate enough lift for flight, a plane needs a minimum airspeed. If it starts its takeoff roll into a headwind, then the air rushes over the wings faster than the wheels move over the ground—the airspeed is higher than the ground speed—and so it can take off when its ground speed is still rather low. If it takes off into a tail-wind, the situation is reversed, and so a much higher ground speed is needed to reach the required airspeed for liftoff. To get to that higher ground speed requires much more runway, and so there's much less margin if anything goes wrong. The same goes for landing: when flying into the wind, the plane can stay afloat with a much lower ground speed and therefore needs less runway to come to a stop.[8]

But one thing pilots don't keep track of, as it has no effects on either planes or passengers, is *space speed*—speed relative to the universe itself. There's no such thing. The whole idea is meaningless.

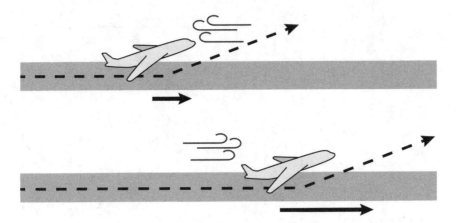

Figure 4: (Top) A plane flying into a headwind can take off with a ground speed lower than its airspeed. (Bottom) A plane with a tailwind requires a ground speed higher than its airspeed and thus more runway for its takeoff.

3

Coasting

Easier Than It Appears

Where am I? And where am I going?

Existential questions such as these pop up repeatedly in life as we make our way through the inevitable troubles of human affairs. We usually ask them metaphorically. But if we take them as concrete, serious questions, we find that they cannot be answered.

At best, they can be partially addressed by referring our position and motion to other objects in the universe, as when we establish where we are on Earth and how fast we move relative to the ground beneath us. When we try to say where the Earth is and how fast it travels, we are led to explain this relative to the Sun or even to our galaxy's center. We can then pinpoint our galaxy's position relative to other galaxies. To finish the job, we would need to say where those other galaxies are and determine their speed and direction. But although we can say where they are relative to one another, we cannot say where they truly are in space. There's no grid crossing the universe, no array of cosmic streets, that would allow us to state or even define the spatial address of a galaxy, or of anything else.[1]

The fact that our universe lacks unambiguous notions of location, both in position and in time, is central to *why* we're poly/omni/ambimotional. Suppose we could measure our current position in the universe, independent of any other object, and suppose, one minute from now, we could similarly measure our new position. Comparing the two positions, we would know how far we'd moved in that minute, revealing our motion's speed and direction across the universe without reference to any other object. Such knowledge would contradict Galileo's principle, which claims that we can't ever tell, if our motion is steady, what our speed and direction might be. Thus, any definite concept of location in our cosmos is forbidden by the principle of relativity itself, and steady motion can be specified only relative to other objects, whose position and motion are equally unspecified except relative to us or yet another object, and so on.

Like most children, I grew up implicitly assuming that time, position, and speed can all be meaningfully ascertained using clocks, maps, and speedometers. It was disorienting to realize that this isn't true. In our universe, there's no place to anchor. We will spend our entire lives unable to state conclusively where we are or where we are going, and that's something we simply have to accept.

3.1 How Relativity Shaped the Modern World

The principle of relativity has played a significant role in human history and culture, at least in some parts of the world. That's a bold claim, so let me try to justify it.

If you watch the Sun and the stars in their daily cycle, rising in the east and setting in the west, it naively seems clear that everything in the heavens circles the Earth once each day. The Earth, meanwhile, appears to be stationary, unique among all things in the cosmos. And that puts the Earth—and human beings—at the dead center of the universe. It's just common sense.

For millennia, the centrality and motionlessness of the Earth were considered self-evident in many cultures. How important we seemed! How wrong we were! Once it's clear that steady motion is undetectable, the slide down the slippery slope has begun.

Quite a few people across history are known to have suggested that the Earth is spinning. Among them were Heraclides of Pontus from ancient

Greece, Aryabhata from the Gupta empire of India, and Abu Sa'id al-Sijzi from Iran a thousand years ago; there were probably many more. Then there was Nicolaus Copernicus five centuries ago, at the dawn of modern European science. But all risked being ridiculed for having no common sense. If we're carried along at hundreds of miles an hour by the Earth's rotation,[2] why don't we feel it? Why doesn't it make us dizzy? Worse, why don't we go flying off the Earth?

These are fair questions. If you place some grains of rice on a spinning plate, the rice will go flying off in all directions. A spinning Earth, it seems, should do the same to us. So serious were these objections that even as late as 1600, long after Copernicus's death in 1543, his view of the solar system was rejected by many astronomers, including Kepler's employer Tycho Brahe. Though convinced by Copernicus and by his own precise observations that the other planets orbit the Sun, Brahe believed that the Sun and stars circle the Earth daily and that the Earth is stationary. "Such a fast motion," he wrote, "could not belong to the Earth, a body very heavy and dense and opaque." His viewpoint had merit: by denying that the Earth moves, he explained why we don't feel its motion.

But in fact, these questions have answers. Today we know that gravity's pull toward the Earth's center is far stronger than needed to counter our tendency to go flying. (In the same sense, if you put something sticky on the spinning plate and embedded the rice grains in it, they'd no longer fly off the plate so readily; to shake them loose, you'd have to rotate the plate at very high speed.) Were the Earth's spin so rapid that a day lasted just a few hours, a person near the equator would have significantly looser ties to our planet than a person at midlatitudes. With our languid twenty-four-hour day, however, any latitude-dependent consequences are too small to notice; they are detectable only with precise measurements.

Furthermore, as I pointed out earlier, our rapid motion around the Earth is at a constant speed and in a nearly straight line, one that curves by only one degree every four minutes. Thanks to the principle of relativity, such near-steady motion is nearly undetectable, and that's why we don't feel it.

Unfortunately, these answers are much more subtle than basic common sense and weren't available in pre-Newtonian days. Since no one yet had a complete understanding of either gravity or motion, it was difficult to have a conclusive debate about these issues.

A spinning Earth could still, in principle, be located at the center of the universe. But it's not so; as Kepler's precise measurements confirmed, our planet circles the Sun at 20 miles per second. This faster, steadier motion also goes unnoticed thanks to Galileo's principle.[3]

Once we recognized that Earth is in orbit, we knew the cosmic center lies elsewhere. Yet the Sun still seemed unique, outshining all other lights in the heavens, and all the other planets orbit it, too, as if it is the hub of the universe. By virtue of our proximity to this dominant and central source of light and heat, we still claimed special status.

But oh, the fall that was to come. Even in classical Greek times, it had been considered that the Sun might just be an ordinary star, viewed close up. The idea received scientific scrutiny from the seventeenth century onward and was confirmed in the nineteenth. Shortly after the turn of the twentieth century, it became clear that we live within a giant star-city, a megalopolis of suns—a galaxy. Galileo's telescope had already revealed that the Milky Way, the white band crossing our night skies, is made of a myriad of glowing points of light; you can check this yourself with good binoculars. Today we know that it is an edge-on view of our home galaxy, an immense spiral-shaped cloud of seemingly innumerable stars.

Within this vast city, the Sun is an unremarkable star, a bit above average in some respects but certainly not extraordinary. It is located far outside our galaxy's dense urban core, relegated to the quiet suburbs, the realm of its looping spiral arms.

The Milky Way galaxy, fairly large but still rather ordinary, is one of many billions scattered across a gigantic expanding universe, and there's no evidence that it lies anywhere near the universe's center. In fact, no such center seems to exist. The galaxies move across the emptiness at great speeds relative to one another. The Sun, the Earth, and the human species are carried along, oblivious to the big picture.

In fact, we have no idea how big the picture is, either in space or in time. Scientists often say that the universe is nearly fourteen billion years old and describe it as though we can see all of it. I will do so myself in this book. But by "universe," they and I are really referring only to the region of the universe that we can observe with our many types of telescopes, which would more accurately be called the *visible universe* or the *known universe*. This may be just one small patch in a vastly larger and/or older cosmos, one whose totality is far grander in scope. We can only speculate as to whether there might exist other cosmic patches, much farther away or more ancient, and whether their basic laws of nature might be completely different from those of our own. (A universe with a patchwork of laws is sometimes called a multiverse, though I won't use the term in this book.) Do keep these limitations on our knowledge in the back of your mind.

The undetectability of steady motion made it seem that human beings live at the stable, central core of the cosmos. It supported the naive intuition that our existence is fundamentally tied to the universe's creation. The discovery of the principle of relativity helped us learn humility, revealing to us that we live in the middle of nowhere and aren't near the center of anything. We wander an immense void at great but pointless speed, with no destination. It's no longer easy to argue that Earth and its creatures are uniquely special.

Other discoveries in science have given us additional perspective on our place in the universe. Recently we've learned that many stars are accompanied by multiple planets; our Sun is not unusual in this regard. The biology of all large Earth organisms is based on similar biochemical molecules, indicating that we're just one species among many, with a common history. Meanwhile, the intelligence and emotional depth of whales, dolphins, elephants, and chimpanzees aren't as limited, compared to ours, as many used to think. Every day it becomes harder to believe that we're the only creatures in the whole cosmos capable of intricate language, abstract thought, and complex emotion.

But in answering the questions of Brahe and other skeptics, the principle of relativity helped to settle the debate over Copernicus's proposal and to open our eyes to the potential vastness and changeability of the cosmos. These realizations permanently dislodged us from our self-appointed throne, our supremacy over a small, illusory kingdom founded on common sense. Looking back, it's all rather embarrassing.

The principle of relativity played a pivotal role in the development of modern science, too. Decades after Galileo first recognized it, Newton took it as the foundation of his principles of motion and of gravity, including his three "laws of motion"—i.e., rules by which the world seems to operate. The second and third laws have been revised over time. But the first, often known as the law of inertia and proposed already by Galileo and even by several earlier thinkers, has survived several centuries of intense scientific investigation. Because the word *inertia* has multiple meanings, it leads to confusion, and so I will not use it here; instead, I will refer to this law as the *coasting law*.

Tightly intertwined with the principle of relativity, the coasting law asserts the following. *An object, if moving steadily and if left on its own (specifically, if unaffected by any push or pull created by other objects), will coast forever; it will continue to move steadily at the same speed and in the same direction.*

This statement includes objects that, from your perspective, aren't moving at all. In other words, a stationary object left alone will remain stationary.

As stated, this law might seem to contradict what we know about animals and about machines with engines. We humans needn't remain stationary— we can just decide to start walking—and a spaceship can use a rocket engine to accelerate without any external push. But in fact, these situations violate the premise of the coasting law, which applies only to a single isolated object (or an isolated collection of objects). A walking creature is not "on its own"—it's on the Earth and pushing its feet against it. And a spaceship with an engine is not a single isolated object, either; the ship and the exhaust from its engine consist of multiple components, and the coasting law therefore cannot be applied to the ship alone.

The coasting law confuses many students who encounter it in a first-year physics class because it runs afoul of our ordinary intuition. Just ask any child. If you'd told me about this law at a young age, before I started reading about physics, I'd have insisted that it can't be right. "Coast? That's just silly! Everything comes to a stop eventually!"

It's just common sense. Throw a ball. Smash a glass. Sweep dust across the floor. First, there's motion; then, after a little while, it ceases.

But as Newton explained to the rest of us, the main reason nothing coasts on Earth is because of *friction*, the rubbing of one object against another. Often this friction creates *drag*, a force that acts as though it were "trying" to keep objects that are in contact from moving past each other. Every object you see in daily life is subject to friction, and so to understand why the coasting law is true requires moving beyond your daily experience. It requires imagination. You have to imagine how things would behave in the absence of friction.

For instance, take a book and slide it across a table. The book scrapes against the table, and the rubbing of their surfaces causes drag, which slows the book's motion until it comes to rest on the tabletop. But now imagine taking the same book and sliding it across a frozen pond. The book will travel farther, and slow down more gradually, because the ice is slippery and causes much less drag than a typical table. With a thin layer of water or oil on the ice, there's even less friction. The slicker the surface, the more gradually the book slows and the farther it can go.

If we affixed magnets to the underside of the book and put it on a surface made of a special material called a superconductor, the superconductivity of the surface would cause the magnets to levitate and the book to float. If we then pushed the book, it would glide over the surface without touching it and would experience no drag. Instead, it would coast, just as Newton claimed, until it reached the edge of the superconducting surface.[4]

What we need to do, then, is imagine what the world would be like if every surface were infinitely slippery and if there were no air resistance. Then we begin to see the world as Newton understood it. Now if someone pushes you, you'll coast through the room, unable to stop until you hit a chair or wall. You mustn't bump your dinner plate, as it will drift across the table and over the edge. Pieces of a shattered glass bottle can easily slide across an entire parking lot. Don't try to walk normally; if you try to stride on a frictionless surface, you'll go nowhere, like a person flailing hopelessly on an icy sidewalk. Only with thoughts like these can we discern how friction dominates our lives and forms the foundation for our common sense about motion.

Long before Galileo and Newton, there was Aristotle's law of motion, or the resting law. This law, which was once widely viewed as obvious, asserts that the natural state of all solid objects is to be stationary—i.e., at rest. According to Aristotle, moving objects, left alone, will slow down and come to a stop. An object can continue in steady motion only if it is being pushed, perhaps by a person or by an engine.

Nowadays, with our understanding of how moons, planets, and stars move, we know that the resting law can't apply to everything in the universe. If it did, the Sun, Earth, and Moon would all slow down. As this happened, gravity would pull them together; first the Moon would crash into the Earth, and then their molten remnants would disappear into the Sun. In fact, all planets and moons in our solar system, and those around all stars in the universe, would suffer the same fate. Finally, the stars would fall into their galaxies' centers. Gravity would rule the universe, destroying everything in it.

This hasn't happened—there's not the slightest sign of it—so observation disproves the resting law as far as outer space is concerned. It's more challenging to prove that it's false on Earth. But fundamentally, the resting law suffers from a deep conceptual problem: it conflicts with the relativity principle. In a universe that operates according to Galileo's principle, the coasting law must be true and the resting law false. Here's why.

The coasting law and the resting law agree that stationary objects will remain stationary. But about moving objects, they differ: the coasting law says they coast, while the resting law says they decelerate and stop. Therefore, if a moving object declares that it wants to travel forever in steady motion, the coasting law smiles and says, "No problem at all; just make yourself comfortable, and go read a book or take a nap." But the resting law frowns and shakes its head, saying, "Hmmm, that will be expensive. You'll need an engine and an inexhaustible supply of fuel, unless you can convince someone with infinite stamina to push you."

In this way, the resting law insists that being in steady motion is fundamentally different and distinguishable from being stationary. This conflicts with the principle of relativity. Something has to give: Aristotle's law or Galileo's principle.

By contrast, the coasting law coexists comfortably with Galileo's principle. In the absence of complicating effects such as friction, both steady motion and no motion are equally effortless; there's no observable difference between them. Neither needs an engine or an external push.

Over nearly four hundred years, these notions of coasting and relativity have remained intact and intertwined. They survived even the great scientific revolutions of the early twentieth century, when Einstein and his colleagues revised our notions of space, time, and gravity, turning much of physics on its ear.

Yet our common sense has trouble with the coasting law, because we never see ordinary objects coast. Our everyday motion, whether by foot, bike, or car, never comes for free; without exertion or fuel consumption, we soon come to a halt.

Experiencing the effortless nature of motion is easier outside Earth's atmosphere. Though jogging a few miles on Earth's surface will leave you sweating and out of breath, astronauts on space walks outside the International Space Station can cover that distance every second, relative to Earth's surface, without laboring at all. They float quietly in the airless regions above our planet, all while traversing oceans and continents in minutes. To cross those continents within the atmosphere, in a jet aircraft fighting air resistance all the way, requires engines and lots of fuel (Fig. 5). Similarly, a car on Earth can travel only a few hours and a few hundred miles before needing to be refueled or recharged, but a car launched into space can cruise in wide loops around the Sun, like any planet or asteroid, for billions of years.

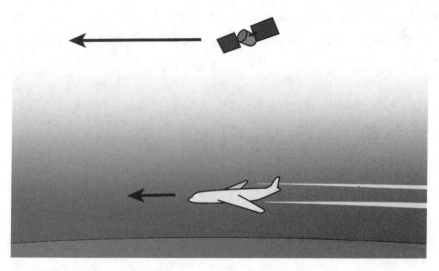

Figure 5: An airplane, flying through the atmosphere, must run its engines continuously to fight air resistance. But a satellite above the atmosphere can coast at much higher speeds than a plane without using an engine.

The Earth, too, glides easily through empty space, as there's no friction that could keep it from coasting. It needs no engine, no fuel, no friendly giant to keep it in its orbit. For this we should be grateful.

I've been referring repeatedly to *empty space*, but I haven't really defined it. You might wonder whether it's the same as *outer space*, which I've also referred to. But the two ideas are different, even though there's a lot of pretty empty space in outer space.

In many contexts, *outer space* means "far enough away from the Earth to be outside its atmosphere." In others, it refers to any part of the universe far from all stars, planets, moons, and other rocky things, which is also sometimes called *deep space*. (NASA brings the term "deep space" closer to home, but I'll use it, as astronomers often do, to mean intergalactic space— the exceptionally empty regions between galaxies.) But while it's true that outer space is mostly empty, and deep space even more so, what I mean by *empty space* is a region of the universe that has been made as empty as it can be.

If you have a box, and you remove everything from it that can possibly be removed—all the atoms and subatomic particles of all sorts—what you're left with inside the box is empty space. It's also sometimes called *the vacuum*. It's the closest thing to nothingness. Yet it has turned out to be remarkably interesting, as we will see later in this book, to an extent that would have surprised physicists before Einstein.

3.2 The Phib and the Principle

Now that we are well armed with the principle of relativity, let's return to the Higgs phib. Even without knowing what mass is or what fields are, we can already appreciate that the tale can't be right. Here it is again, in a more elaborate version.

Once upon a time, at the earliest moments of the universe, the Higgs field slept. Lacking mass, objects zoomed rapidly and aimlessly about, alone and glum.

But then the Higgs field woke. It filled the universe from end to end, from side to side, from top to bottom, and forevermore.

Ever since that moment, the Higgs field has surrounded us like a vast sea of molasses—or, in other versions of the tale, like an endless soup, a dense thicket, a great crowd of people, or a thick blanket of snow. As objects try to move through the Higgs field, it slows them down. By slowing them down, it gives them mass; the more it impedes their motion, the more mass they acquire.

That's how objects in the universe were endowed with the mass they have today. Able now to congregate, they made new friendships and began to dance together around the cosmos. Soon, as the sky filled with stars, the universe became the one we know.

For a moment, the phib might sound plausible. But it's a swindle. By asserting that objects with mass do not coast but instead slow down, it contradicts the coasting law. It violates the principle of relativity, too, since it claims that the Higgs field would do one thing to steadily moving objects (it would act to slow them down) and a different thing to stationary objects (nothing at all). If this were true, it would require that the Higgs field do something deemed impossible by the principle of relativity: distinguish steady motion from no motion.

You, too, would be able to determine your motion, even inside an isolated bubble. The phib alleges that because you and the bubble have mass, the Higgs field would slow you down if you were moving. But you can detect slowing down even in an isolated bubble. You know this from experience: when the car or plane that you're in decelerates quickly, you feel as though you're being pushed forward in your seat. So if you were moving and the Higgs field were slowing your motion, you'd feel it. In contrast, if you were stationary, there'd be no slowing, and you'd feel nothing. The difference would allow you to determine easily whether you were moving or not. That would be completely incompatible with Galileo's principle.

Fortunately, the Higgs field isn't and never was in the business of slowing things down. Had it been, then the resting law would have been true, with all its catastrophic consequences. All objects would have ground to a halt thanks to the Higgs field.

Back when the resting law was widely believed, and before anyone understood the nature of the Moon, Sun, or planets, many protoscientists did think slowing and mass were connected. But in truth, slowing things down doesn't give them mass, and mass doesn't make things slow down. Steady motion, at a sprint or at a crawl, is just as acceptable for giant planets as for subatomic particles. And all objects, no matter what their mass or speed, will coast if left alone.

The Higgs phib is thus flagrantly inconsistent with Galileo, Newton, and Einstein. It touts a premodern view of the cosmos and injects a profound internal contradiction into the heart of physics. Only by abandoning it can we hope to maintain a coherent conception of the universe, preserving the principle of relativity at its core.

It's much easier to cast shade than to shed light. Though I've quickly exposed the problems with the Higgs phib, to replace it with something honest and comprehensible will take much longer. But before this book comes to an end, we will see how the Higgs field and the relativity principle can live in harmony.

MASS

According to the principle of relativity and the coasting law, continuing in steady motion at any speed requires no effort. But if it's that easy, we might well wonder why there isn't more motion in daily life, even outside modern vehicles. Why don't our possessions fly about our bedrooms, and what limits us to walking and running speeds relative to our surroundings?

There is plenty of high velocity around us; we're just not aware of it. The air molecules currently bombarding your face are traveling (relative to you) faster than a jet aircraft. Still, they are slow compared to the occasional particle that comes flying out of the walls of the room, or even from deep within your body. Such particles are created by a natural process called radioactivity, in which an atom is transformed from one type to another. Radioactivity in large amounts, such as those generated by nuclear power plant accidents, can be very dangerous. But even in the cleanest environments, there's always a small quantity of radioactivity within and around us—indeed, natural radioactivity helps provide Earth's internal heat, keeping the continents adrift—and so its agile by-products are inescapable. Fortunately, our bodies can repair the minor damage that they cause.

Meanwhile, every few seconds a particle called a *muon* (usually pronounced MYU-on) passes through your body, having been created far above your head. The origin of this muon is a *cosmic ray*, a generic term that refers to any high-speed particle from outer space. When a cosmic ray crashes into an atom in the Earth's upper atmosphere, muons produced in that collision careen toward Earth's surface at breakneck speeds, a million times faster than a passenger jet.

You might think it strange that we don't feel these speedy demons impacting or traversing our bodies. But knowing about them wouldn't provide us with any clear advantage as far as surviving and reproducing; they offer no clues as to the location of food, danger, or potential mates. Consequently, nothing during evolution has pushed humans or other animals to develop nerves sensitive to their presence. Not that it would have been impossible—our eyes can detect the extremely microscopic, as can our sense of smell. Yet every new sense comes at a price; it has to be built as a fetus develops, and requires resources. If it isn't needed, it's not worth it. This is why our senses, our brains, and our common sense have been left out of the loop.

Too bad for us. But the question remains: With this maelstrom of headlong motion all around us, how do we avoid being sucked into it? Nothing intrinsically inhibits our motion; indeed, we can and do move swiftly relative to the Sun, distant stars, and so on. Friction is part of the answer. Were we suddenly flung violently across a room, air resistance and friction from the ground would quickly slow us back down. But why aren't we ever flung in the first place?

Whatever the answer, it's a good thing. I would not want to enter through my front door at the speed of an average air molecule.

There's a character trait that we humans are known for. As a species, we're stubborn. We're suspicious of change and tend to resist it. This psychological intransigence has some benefits as well as costs. It turns out, though, that it's not merely our brains that are stubborn; our bodies are, too. It's this physical resistance to change, which we call *mass*, that keeps us safe amid the invisible swarming chaos of the natural world.

4

Armor Against the Universe

Let's lead off with a fable.

Imagine our planet hosts two human species: *Homo sapiens* (that's us) and *Homo polystyrene* (our cousins, the Styrenians). Our cousins resemble us in shape and size, and at first glance it's hard to tell us apart. But internally, they are different. Their flesh and bones are spongy and lightweight, as flimsy as the packing material known as polystyrene, sold widely as Styrofoam. You could easily carry an adult of their species, who will have no more heft than you did as a baby.

You might imagine that our Styrenian cousins are the lucky ones; they are lighter on their feet and perhaps can get around more quickly. But then imagine life on a day with a breeze.

On a windy day, you might have to hold on to your hat to keep it from flying off. But if you were a Styrenian, that would be the least of your problems (Fig. 6). Even as an adult, you'd have to hold on to a wall or post to avoid being hurled high into a tree or flung into a river. Daily existence would be hazardous, akin to living in a hurricane.

Mass is your armor against the world, allowing you to resist nature's efforts to blow or sweep you away. With the same size but much less mass, you'd be tossed around like a plastic bag or a dead leaf. A falling twig could knock you flat. You might even struggle against the gentle push of sunlight.

Figure 6: A member of our species might lose a hat on a windy
day, while a Styrenian, with the same size but far less mass,
might lose contact with the ground.

So as you walk without a care in the world down a country road or city sidewalk, do not take your mass for granted. The life of a Styrenian is not to be coveted.

The property called *mass* is a form of tenaciousness, of stubbornness, in the face of change. Specifically, an object's mass is an intransigence that hinders any attempt to alter the object's motion—to cause the object to speed up, slow down, or change direction. If the object is stationary, its mass inhibits any endeavor to make it move.

The wind blows and tries to toss you across the street. Your body resists this effort, but not because of its size or height, its thickness or chemical makeup, its age or temperature. It resists because of its mass.

We're all familiar with this stubbornness of objects. Rocks don't make it easy for you to throw them. To heave a stone over a creek requires exertion. The greater the mass of the stone, the more exertion is needed.

If an object is already moving, it resists your attempts to change its speed or direction. (This is why *inertia* is sometimes used as a synonym for *mass*, though I'm avoiding that confusing term in this book.) Suppose I toss a fist-sized ball toward you, and you want to catch it. It's easy to slow it to a stop with your bare hands if it's a hollow plastic sphere, but if it's a standard baseball, with much more mass, you'll have to work a lot harder. And if it's made of solid lead, beware!

This last example illustrates that mass isn't related to size unless you are comparing two objects made from the same material. It's true that two bricks have more mass than just one and that a large granite rock is more intransigent than a small one. Nevertheless, that same granite rock has more mass than a bottle of water twice its volume, and your Styrenian twin sibling has far less mass than you do.

4.1 Figures of Speech

For a complete understanding of mass, it is crucial to recognize what it is not: matter. The relationship between these two words often causes confusion and serves as an example of how details of human language can obstruct comprehension of the universe. Such linguistic concerns might seem out of place, or at best tangential, in a book on physics. But they are not.

Language plays a central role in any collective human endeavor, and science, despite its reputation for rigorous methods and mathematical laws, is hardly immune to the challenges of human communication and miscommunication. The fact that scientific terms are often unfamiliar or, worse, familiar-seeming poses a potential problem in any book like this one.

The language spoken among particle physicists is similar to English in many ways, but it is a dialect of its own, blending ordinary language, scientific jargon, and something more insidious: false friends. A false friend is a word in another language or dialect that sounds just like one you know, making it seem as though it needs no special definition. An example is the word *médecin* in French; most English speakers would naturally guess that it has the same meaning as *medicine* in English, but in fact, it means *doctor*. In physics dialect, this is a common issue. Simple terms such as *mass, matter, force, wave, field, value*—come to think of it, almost every important word used in this book—have meanings that differ from their more standard meanings in ordinary English. Technical jargon such as *adiabaticity* or *thermodynamics* is actually less of a problem, as such words clearly need to be defined before they are used. False friends are the most dangerous; they seem kindly at first, but they will stab you in the back. I am afraid that we will have few true friends on this journey, so be on your guard (and if need be, make ample use of this book's glossary!).

Not that this is surprising. Every technical subject has a similar dialect with its own false friends. Followers of baseball know that *walk, run, strike,* and *base* have new meanings within the game, whereas *love* means something quite unexpected in tennis. Musicians, for whom *sharp* and *flat* have nothing to do with shapes, redefine many other words, including *scale, measure,* and *tonic.* In weather, a hurricane can't see even though it has an *eye,* and there's no warm back behind a *cold front.* In almost any field of human activity, language is partially recycled; words that are widely understood in English are repurposed and given unfamiliar definitions.

If you're well versed in a subject, it's all too easy to use a false friend without remembering that it carries a novel meaning. I speak from experience. Consider the sentence "Electrons are massive particles." In English, this claim is bizarre; an electron is exceedingly tiny, so the sentence seems ridiculous. But in physics dialect, where *massive* is redefined to mean "possessing nonzero mass," this sentence simply indicates, correctly, that "electrons have mass." I'd been writing blog posts with similar sentences for six months before a thoughtful reader pointed out that *massive* is a false friend. Since then I have carefully avoided using the word where it could be misinterpreted, and it does not appear again in this book.

I similarly avoid the word *matter* whenever possible, but here the problem is a little different. Tragicomically, words are often given diverse and contradictory meanings in different areas of science and math, with the result that communication sometimes breaks down even among experts. For instance, the word *field* means one thing to physicists and something completely different to mathematicians, and neither is what it means to farmers. Would you believe that Earth's air is made of *metals?* For many physicists, a metal is a solid crystalline material that conducts electricity, just as in English, but astronomers use that word for any atom whose nucleus was forged inside a star. By that definition, which includes every element except hydrogen and helium, not only Earth's atmosphere but also its living beings are all predominantly made of metals. When you next look at the metallic creature in the mirror, consider how words and definitions affect the way we think.

Few words are worse, in this regard, than *matter*. It has at least two contradictory definitions used by particle physicists and at least two others used by astronomers. All the definitions agree that atoms (and thus all ordinary objects) are examples of matter, while light is not. But they disagree as to whether more exotic objects, such as Higgs bosons, neutrinos, and antiprotons, are matter. It's not even clear that "dark matter," a term used widely by astronomers and particle physicists alike, is actually matter.[1] So ambiguous and potentially confusing is *matter* that I will bend over backward not to use it; instead, I'll usually refer to objects made from atoms as "ordinary material."

Yet another problem is the shorthand often found in scientific dialects. To keep the language from becoming long and tortuous, complicated ideas are often compressed, abbreviated to just one or two words. An example of this is what scientists call "dark energy." Long ago, the universe expanded rapidly, but the rate of its expansion has slowed over time. Recently, however, the universe's dark energy has been preventing it from slowing further. But dark energy is not energy! It's something—perhaps a field similar to the Higgs field, or perhaps an aspect of empty space itself, or a combination thereof—that *has* energy (along with negative pressure). True though the last sentence may be, it is both vague and long-winded. That's why everyone uses the two-word label; it's convenient, brief, evocative, and easy to remember. The fact that it's not accurate doesn't bother the experts because they know what's hidden. The label is less benign when used in conversations with nonexperts.

For the opening sections of this book, the most important thing to remember is the following: matter is a substance out of which objects can be made, whereas mass and energy are properties of objects, not substances. More generally, it will be crucial in this book to distinguish substances from properties.

As we saw in the Styrenian fable, mass is the property that reflects how well an object can resist the power of the wind, the impact of a running child, or any other of the slings and arrows of existence that try to push it around. It's one of many properties that your body has: height, thickness, shape, temperature, strength, age, and so on. In contrast to matter, none of these are substances; you can't build objects out of height or age, and you can't build them out of mass or out of energy, either.[2] Instead, ordinary objects are all built out of atoms, a form of ordinary matter. Unfamiliar objects can potentially be made from other substances.

Regarding mass and matter, there's another issue raised by what many of us were taught as schoolchildren. In my chemistry class, I learned that the mass of an object is the "quantity of matter" inside it. It's a definition that was introduced by Newton himself. He argued that the more matter an object is made from, the more difficult it is to change its motion. We will soon see why this viewpoint works well in chemistry class and in daily life.

But in subatomic physics, serious flaws appear. We'll soon encounter objects that have mass but are not (by most definitions) made from matter. Other objects that are clearly matter (by all definitions) may in the past have had no mass whatsoever. These are not the only problems.[3] The definition of mass used in modern physics does not refer to *matter* at all.

As we proceed through this book, these linguistic threats to communication will arise repeatedly. In pointing them out, I suppose I risk coming across as a tedious grammarian, an antiquated curmudgeon who complains that nobody uses words properly anymore. But I think you'll soon see why this attention to what we say (and don't say), and to what our use of language hides from view, is crucial for an understanding of the cosmos. The incoherence and inconsistencies of scientific discourse serve as a window into how we humans think and into how science is actually done. By looking through that window, we can reduce our reliance on the imperfect words we use to describe the cosmos, helping us perceive its essence more clearly.

4.2 Weight and Mass

For thousands of years, natural philosophers assumed that an object's weight and mass were the same thing. But in fact, they're distinct both in principle and in practice, as Newton himself realized.

An object's weight represents how heavy it feels when you hold it steadily off the ground, defying the pull of gravity. But an object's mass is an intransigence that makes it difficult for you to throw it or catch it. (It's important that the throw be horizontal in order not to mix up the effects of weight and intransigence.) It's far from obvious that these two different activities of supporting and throwing, shown in Fig. 7, should have anything to do with one another (and in some universes one can imagine, they wouldn't). In daily life on Earth, however, we find they are always related: the harder it is to hold an object up, the harder it is to fling it. We learn this in early childhood; more weight means more mass. It's almost common sense that the two words are redundant names for the same thing.

These observations aren't limited to throwing. Any other method of altering an object's motion reveals the same thing. Just as shouldering a box of

Figure 7: The difficulty of holding an object against gravity's pull (gray arrow) grows with its weight, while the difficulty of throwing an object horizontally (black arrow) grows with its mass. Out in deep space, the object would have no weight, but its mass would be the same; it would be as difficult to throw as ever.

bricks requires a lot more strength than shouldering a similarly sized box of packing foam, changing the former's speed in any way is much more difficult. For instance, suppose you placed the two boxes on lightweight wheeled carts (to reduce the influence of friction) and then gave the two carts an equally firm shove. Because the bricks' greater intransigence would more effectively resist your push, their cart would end up moving much more slowly than the cart with the foam.

The apparent identity of mass and weight has an important consequence, which you can witness in a do-it-yourself experiment. Take several metal objects with various weights: maybe your keys and two very different coins. Also take two books, one very heavy and the other much lighter. Hold them all in your two hands (or have a friend help you) at waist height. Then, at the same moment, drop them all.

If you haven't ever performed this experiment yourself and have only read about it or watched it done, I encourage you to try it. Every now and then I carry it out for a friend or acquaintance, both to renew my appreciation for our remarkable universe and to enjoy the satisfying "wham" created by all the objects landing at exactly the same time. Besides, physics is an experimental science, and even though you might know something amazing is true, there's no substitute for seeing it with your own eyes.

I recently demonstrated this to two seven-year-olds. The looks on their faces were priceless; the boy was especially wide-eyed. The girl cried, "Magic!" to her mom; she thought I was performing a trick, like a stage magician, until she tried it herself. Indeed, it is rather magical to see a paper clip and a heavy rock do exactly the same thing, as though they were rigidly connected by an invisible rod. As a friend of mine remarked, physics and magic are both mysterious, but magic is illusion, while physics is its opposite.

Most children, having seen feathers and pieces of paper glide to the ground more slowly than rocks fall, develop an intuition that heavy objects fall faster than light ones. Rarely do they notice on their own that this intuition is false, and unless their science class makes an impression, they may neither question common sense nor notice that it disagrees with reality. Feathers and paper mislead us because they are strongly affected by the air they are descending through; the air resists the effects of gravity's pull. Objects less subject to air resistance, such as a pebble or a metal paper clip, plummet in the same way as a large rock. As long as a small object's motion is dominated by gravity, its descent once released is independent of all its properties, including its shape, its size, and even its mass.[4]

The reason objects fall in lockstep—why they accelerate at the same rate as they fall toward the ground—is that at the Earth's surface, every object's mass and weight are related to the others' in exactly the same way. This is why we can infer an object's mass by measuring its weight, as long as we stay here on Earth.

For instance, on Earth a nickel weighs twice as much as a penny—gravity pulls harder on the former than on the latter. But the nickel also has twice as much intransigence as a penny, so it's half as responsive to that gravitational force. Since the nickel's larger mass exactly compensates for its larger weight, the two coins pick up speed in exactly the same way.

As long as we're on the Earth's surface (or pretty close to it, as in a passenger airplane or submarine), it's basically true that mass and weight are interchangeable. But Newton realized that there are situations in which gravity can change but intransigence does not—in which an object's weight becomes larger or smaller but its mass remains fixed. Specifically, what an object weighs depends on where other objects around it are located, which is not true of its intransigence. That's why Newton gave intransigence its own name: *mass*, distinct from *weight*.

For Newton, mass is one of an object's intrinsic properties, something it possesses all on its own. The pull of gravity, on the other hand, stems from a relationship between multiple objects. As an object moves around the cosmos (Fig. 8), its weight changes but its mass does not.

As you travel the universe, the pull of gravity on your body will vary substantially, depending greatly on how far you are from the Earth, Sun, Moon, and so on. On Earth's surface, we are roughly 4,000 miles from Earth's center. But if you ascended another 22,000 miles, where you'd find the GOES weather satellites[5] that monitor Earth's weather patterns, you'd find your weight (but not your mass!) reduced to one-fortieth of what it is on Earth, comparable to the weight of a pillow in your bedroom. On the Moon, where the Earth's pull would be tiny, the Moon would pull you toward its center with gravity that's about one-sixth of what you're accustomed to. And if you traveled out into deep space, far from any large object, you'd weigh virtually nothing. Yet all the while, your body's mass—the difficulty I would face if I tried to speed you up or slow you down—would never change.[6]

When I explained this to a friend, he asked, "Hmm—so even though it would be much easier to *hold* a box of bricks on the Moon than on Earth, it would be just as hard to *throw* it?"

"That's right!" I affirmed. "As hard to throw or catch as it is on Earth. Although the box of bricks would weigh one-sixth of what it weighs here, would fall much more slowly if you dropped it, and would be easier to carry, it would hurt you just as much if you accidentally ran into it."

For the same reason, there's no free lunch waiting for you in the blank spaces between the distant stars. Gravity is so feeble there that your spacecraft and its contents would be almost weightless. Were they almost massless, too—if they lacked their usual intransigence—then they'd be dramatically more responsive to a push than they would be on Earth. A moment's firing of your rocket, using almost no fuel, would accelerate you and your craft to blistering speed, rushing toward home. Sadly, your spaceship's intransigence would not, in fact, have changed. The initial rocket burn needed to send you quickly earthward would require a great deal of fuel.

To see this another way, let's have a friend take a nickel 22,000 miles overhead, the altitude of the GOES weather satellites. Our friend will drop this nickel at the same moment that we, back on Earth, drop a penny and a nickel from waist height. As always, the earthbound penny and nickel will fall together and hit the floor in a fraction of a second. But in that time, the second nickel will fall less than the length of your thumb. Its slower fall reflects the difference between weight and mass. The two nickels have the same mass, but so far from Earth's center (and even farther from the Moon and Sun), the second nickel has a greatly diminished weight. In other words, although its intransigence is the same as that of the earthbound nickel, the gravitational pull to which it is subject is much weaker. Consequently, it falls much more slowly than do the two coins near Earth's surface.[7]

Some years ago, after a class about these topics, a few students gathered around me asking questions. After I explained how gravity's strength changes during a trip from Earth's surface to the Moon's, one of the students, eyes twinkling, quipped, "Lose weight! Go to the Moon!"

Another retorted, "Yeah, the Moon diet will be a fad someday, until people figure out that what they actually want to lose is *mass.*"

My jaw dropped. I'd never thought to point this out, nor had any of my teachers.

Instantly a third student clapped hand to forehead and cried, "No *wonder* watching my weight never works!"

M

E

G

Figure 8: Your weight varies with your location; your mass does not. Near geostationary satellites (G), your weight (from Earth's pull) would be one-fortieth of what it is on Earth (E). On the Moon (M), where Earth's pull is tiny, the Moon's pull would give you a weight one-sixth of what it is on Earth. Out in deep space, well away from all other objects, your weight would be nearly zero.

If you drop your keys, it takes roughly half a second for them to hit the ground. The Moon orbits the Earth every twenty-eight days. These two facts don't obviously have anything to do with one another. Yet Newton suspected that they are related—that the same gravity force that makes your keys fall also holds the Moon in its orbit. The idea might seem so simple and obvious now that it can be hard to understand why it took humans so long to think of it. But there's a reason. On the face of it, the idea is fatally flawed. It's obviously wrong!

Wrong, that is, if weight and mass are two words for the same thing.

Philosopher-scientists before Newton largely assumed that the rules of the heavens differ from the rules governing objects on Earth. It's common sense. Everything we ever drop falls to the ground, while the Sun, Moon, planets, and stars seem to float in the sky. Perhaps, some thought, they feel no gravity, no pull toward the ground. Others supposed that these luminous objects naturally waft upward, much as flames and smoke rise off burning wood. Still others suggested that the heavenly objects are attached to transparent spheres that rotate around the Earth once a day.

By Newton's time, it was clear that none of these proposals was correct. Instead, it's all about motion of one sort or another. The daily transit across the sky by the Sun and other heavenly bodies isn't true motion; it's merely an illusion due to Earth's rotation. But the Moon's monthly transformation, from full to new and full again, reflects its orbit of the Earth. The changing location of the Sun from winter to summer and the complex paths of the planets as they wander across the sky reveal that the planets, including the Earth, orbit the Sun.

These orbits maintain themselves for eons through a balance between gravity and motion. For the Moon, this is illustrated in Fig. 9. Were there no gravitational pull from the Earth, the Moon would coast, traveling at a constant speed and in a straight line. On the other hand, if the Moon were momentarily stationary relative to the Earth, the force of gravity would pull the Moon straight "down" (i.e., toward the Earth's center). To the extent that the Moon's initial motion is sideways in the figure, its path will curve as it falls to Earth. The faster it moves, the farther the Moon will travel around our planet before a collision occurs. Finally, when it is moving fast enough but not too fast, motion and gravity balance: the Earth can't drag the Moon down, but the Moon can't get away, either, and stays in orbit at a more or less steady distance as it circles the Earth. The same type of balance keeps the planets in orbit around the Sun.[8]

But when Newton first began thinking this might be true, he was well aware of a serious potential obstacle. The Moon's distance from Earth (about 240,000 miles, or 400,000 kilometers) had been measured two thousand years earlier, using tricks of perspective.[9] Also known was its speed around the Earth, about two-thirds of a mile (about one kilometer) each second, since this can be learned by combining the Moon's distance with the time

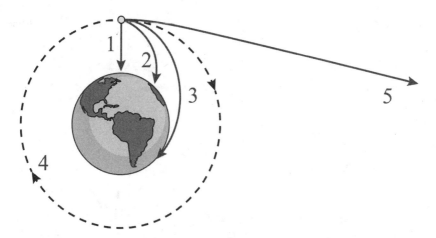

Figure 9: A stationary object, released far beyond the Earth's atmosphere,
would fall toward the Earth's center (1). If moving too slowly to the right, it
would descend and impact the Earth (2), (3). At too high a speed, it would
escape the Earth entirely (5). But at an intermediate speed, its motion
would balance gravity's pull, allowing it to orbit the Earth (4).

it takes to complete one orbit of our planet, roughly a month. So Newton
knew right away that if the force of gravity were as powerful out by the
Moon as it is at Earth's surface—if the Moon accelerated toward the Earth
at the same rate that your dropped keys do—then motion and gravity would
be wildly out of balance.[10] Even as it was first forming, the Moon would have
crashed into the Earth, following a path similar to curve (2) in Fig. 9.

To balance lunar motion with gravity, Newton recognized, the effect of
Earth's gravitational pull must be much weaker out near the Moon than
it is for us here—thousands of times weaker, in fact. Under such a weak
pull, your dropped keys would take thirty seconds to reach the floor! Still,
it's enough to keep the Moon and Earth together. The effect of gravity that
binds the Earth to the Sun isn't much stronger.[11]

Based on the work of earlier scientists, especially Kepler and Huygens,
Newton guessed that if an object's distance from Earth were doubled, its
weight would decrease to one-fourth of what it had previously been. (That's
known as the *inverse square law* because if you multiply the distance by 2,
you can learn the change in the strength of the force by squaring 2 to get 4

and then inverting the 4 to get ¼. Similarly, tripling the distance reduces the weight by one-ninth.) The Moon is roughly sixty times farther from Earth's center than we are, and because 60 squared is 3,600, Earth's pull on the Moon is reduced by about 1/3600 from what it would be if mass and weight were the same. As Newton showed, this is just what's needed to balance the Moon's motion and keep the Moon in its stately orbit.

Newton was just getting started. He boldly proposed that gravity is universal, by which he meant generally that all objects attract all other objects, and specifically that the force between two objects decreases with their separation via an inverse square law, just as Earth's pull on the Moon does. With this in mind, he considered the Sun's pull on the planets. Not only did he correctly predict each planet's speed in its orbit, but he also explained a crucial detail: why the orbital path of each planet is not a perfect circle but instead traces an ellipse.

Then he turned from the sky to the sea. If gravity is truly universal, then just as the Earth pulls on the Moon, the Moon pulls back! Furthermore, because gravity dwindles at greater distances, the Moon's pull is stronger on the near side of the Earth and weaker on the far side than it is on the Earth's center. This uneven pull stretches our planet's oceans slightly, resulting in a small bulge of water, not much taller than a human, both on the Earth's side facing the Moon and on the opposite side, too.[12] As the Earth rotates daily, each location on the ocean's shores thus sees the water rise and fall twice. This was Newton's solution to one of the world's oldest and most mysterious puzzles: our twice-daily ocean tides are created not merely by the Moon's gravitational pull but by the variation of that pull across the Earth. Were weight and mass the same, there would be no such variation and no tides.[13]

Newton's bold idea implies that you pull on your kitchen table, and vice versa. Why don't you feel that pull? Because gravity is a remarkably weak force. The gravitational pull of the entire Earth is too frail to prevent you from lifting your feet or tossing a ball. The pull of your kitchen table, or of any person you might pass on the street, is many billions of times smaller than that.

Newton had an unusual personality. Rather than rushing out to proclaim his discoveries to the world, he kept them to himself for many years. Finally, Edmond Halley, famous for recognizing the seventy-six-year orbital cycle of the comet that bears his name, learned of Newton's extraordinary insights and urged him to make them public. The result was perhaps the greatest book in the history of science, the *Philosophiæ Naturalis Principia Mathematica* (Latin for *Mathematical Principles of Natural Philosophy*), in which the solutions to numerous mysteries of the universe were unveiled for the first time.

Newton thereby transformed a cosmos that most Europeans, and many others around the world, had separated into an earthly realm and a heavenly realm. His successful predictions in so many different settings revealed that there is only one empire, ruled by gravity. The workings of the heavens aren't so mysterious after all. We need not leave the Earth to witness them in action; it's enough to go to the beach.

With that realization—that we are pulled by, and pull with, the same force that keeps the planets and the Moon in their orbits—came a sea change in our view of our place in the universe. Already Copernicus, Kepler, Galileo, and others had established that the Earth isn't the universe's center. From Newton, we learned that the shining lights that traverse our skies are made from the same materials that are found here on Earth. Heavenly bodies and human bodies all follow the same rules.

That's not something we could have guessed by staring at the sky. We needed a Newton to think it all through and to teach us that mass isn't weight.

Once, when I got a bit carried away on this subject, one of my students sat back, stretched, grinned mischievously, and commented, "So it was Newton who made us one with the cosmos?"

I thought about that for a while. The idea that we are one with the universe spiritually has been around for ages. But that we are one with the universe physically—in a completely material sense—is a more radical idea.

Not only did Newton prove that this is true, but he also provided formulas of practical use. Before him, no one could have guessed how to design a rocket trip to the Moon or send a satellite into orbit around the Earth.

Einstein's updates to Newton's formulas are crucial in certain astronomical contexts, including black holes and the expansion of the universe, or when high precision is required, as for GPS navigation and for long interplanetary voyages. But Newton's original methods are still accurate enough for many purposes; even today, space agencies use them in their planning.

Newton's conception that orbits result from a balance between gravity and motion tacitly assumes that there is no friction in empty space that would degrade that motion. His formulas make sense of the heavens only if the relativity principle and coasting law apply there. If instead the planets were slowed by friction, the principle of relativity would be obscured, just as it is for us in daily life. The resting law would apply to planets instead of the coasting law, with dire consequences.

But before Newton's successes, the absence of friction in outer space was far from evident. Because of friction's prevalence on Earth, it is ingrained in common sense, and it would be perfectly natural to assume that it is a universal presence, even in the planetary realm.

Though Copernicus proposed that the planets orbit the Sun, it was Kepler who put the idea on a firm footing. He understood the planets' motions with a precision never previously achieved and even suspected that the Sun exerts a force on the planets. Yet he harbored the same misconception that today leads science fiction writers to attach running engines to coasting spaceships.

Kepler, implicitly imagining that the planets are subject to friction, assumed they would slow and stop unless propelled by a sideways force exerted by the Sun, as depicted by the gray arrow in Fig. 10. This force would push them along in their orbits, much as a parent might nudge a recalcitrant toddler from behind to keep the child moving. It apparently never occurred to him, despite decades of thinking about the problem, that perhaps there was no such friction—that perhaps, in the absence of a force, the planets would not stop and instead would coast, sailing straight ahead as they quickly left the Sun behind them. To remain in orbit, planets free of friction and subject to the coasting law would require a force directed toward the Sun, as indicated by the black arrow in Fig. 10. This was clear to Newton and his contemporaries.

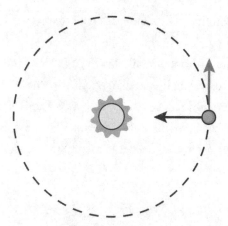

Figure 10: Kepler imagined that planets had to be pushed along in their orbits (gray arrow). Newton, assuming the coasting law, guessed that they had to be pulled toward the Sun (black arrow).

Kepler died, not yet sixty years of age, in 1630. Galileo's writings on the principle of relativity appeared in 1632. Had Kepler lived a few years longer and read Galileo's book, might the bright light of understanding have blazed within his mind?

I tell Kepler's story partly to give us all a break. Few of us are as smart as Kepler was, and even he didn't figure out the principle of relativity or the coasting law, despite staring at the planetary orbit problem and reams of precise data for the entirety of his adult life. For thousands of years, most people, even experts on the subject, assumed that objects can't move steadily unless they are pushed. Over history, a few isolated scholars (notably the Persian thinker Abu 'Ali ibn Sina and French philosopher Jean Buridan) proposed the coasting law, but they didn't manage to incorporate it into a fully coherent vision of how the world works, as Newton did. It took new technology, especially the telescope, combined with a sequence of brilliant scientists following one upon another, for the coasting law to gain universal acceptance and for the variety of motion around us to finally make sense.

Centuries after their discovery, coasting and relativity remain a conceptual challenge. Nothing in our daily lives makes it obvious to us that friction, largely unseen, opposes us at every turn, often requiring us to push an object continually to keep it moving. It's true that we have some advantages over Galileo's compatriots, in that modern transportation offers us more opportunities to gain intuition about relativity than was the case in the preindustrial era. But today's brains aren't superior to those of several centuries or even several millennia ago. What was conceptually difficult for the human mind back then remains difficult now. The most essential difference between our ancestors and ourselves is merely this: they lived before Newton, and we live after.

> *Nature and Nature's laws lay hid in night:*
> *God said, Let Newton be! And all was light.*[14]

5

Enter Einstein
Rest Mass

Many aspects of our universe are relative, including steady motion; we may disagree about them and yet all be correct. But certain features of objects are independent of the observer who views them. These intrinsic properties hold special interest, as they help us to characterize an object's essential nature.

As an example, consider the building where I do research. Its apparent height is relative: if viewed from across town, it looks tiny, while as seen by someone just outside it, it blocks out half the sky. Yet the fact that it is four stories high is beyond debate; anyone who can see it, whether far or close, and no matter what their motion might be, can count the four rows of windows. Four stories is the building's intrinsic height, and it doesn't depend on anyone's perspective.

Many other concepts come in intrinsic and relative versions. One example is brightness. A lightbulb that may seem almost blinding if you're right next to it will appear dim if you're far away. We'd call that *apparent brightness*; it's a relative property that captures how you and the lightbulb are in relation to one another. Nevertheless, the actual amount of light produced, printed on the bulb and its packaging, has nothing to do with you or with anyone else who might be looking at it. It is intrinsic to the bulb, a property of the bulb alone.

Similarly, the loudness of a loudspeaker may intrinsically be set to its maximum volume, but whether it seems loud or soft to you depends on how far away you are. A room intrinsically at 68 degrees Fahrenheit (20 degrees centigrade), as measured by a thermometer, may feel cold on a warm day and warm on a cold day. And so on.

I was once asked, "Why do scientists even bother with relative properties? Isn't it more useful and important to focus on the intrinsic properties of things?"

"Well," I replied, "sometimes scientists want to describe how things look from a particular point of view, or from multiple points of view. For instance, astronomers often need to know both how bright a star appears in the sky from our earthly perspective and how bright it is intrinsically. More generally, to interpret the results of a measurement correctly, understanding the perspective from which the measurement was performed may be critical. So relative concepts are often essential, too."

In Newton's day, mass was thought to be a unique, definite, intrinsic property of an object. It could be determined by measuring an object's intransigence when pushed. Moreover, it equaled the quantity of matter that the object contains. Exactly the same quantity seemed to be responsible for gravity. As far as Newton could tell from the observations and experiments that he could carry out, the gravitational force between two objects is completely determined by the masses of the two objects and the distance between them.

But Einstein guessed, and the technology of his time soon made evident, that different ways of measuring intransigence actually give different answers. When the dust settled, it became clear that mass comes in various versions, and most of these are relative.

Since the concept of *mass* is suddenly becoming more complicated, requiring us to use it with care, I'm going to keep the word *intransigence* around in this book so that we have a term that's more focused on how you do a measurement and less on how you interpret it. The notion is this: if you push on a freely moving object and measure how its motion changes, you're measuring its intransigence as seen from your perspective.

When you push an object that is passing you, you will find, to Newton's surprise, that its intransigence depends on its speed. It also depends on whether you push it in a direction parallel to or perpendicular to its motion. The result of any such measurement is perspective-dependent. A different observer, moving at some other speed relative to the same object and pushing it in the same way as you did, will find its intransigence to be different from what you measured.

We could give names to all the different ways that one can measure intransigence, each of which defines a separate relative notion of mass. But with one exception (called *relativistic mass* and described below), we won't need to name them.

That said, there is one other relative version of mass that we should name, even though its role in this book is minor. Rather than representing a form of intransigence, it is the version of mass associated with creating a gravitational force. Its name, not surprisingly, is *gravitational mass*. It was Einstein who realized, as he developed his understanding of gravity in the 1910s, that both the force with which gravity pulls and the response of objects to gravity are perspective-dependent!

The fact that gravitational mass is relative would be a central element of any book dedicated to Einstein's view of gravity. But in this book, we will concentrate our attention on another crucial version of mass, called *rest mass*.

Unlike the other forms of mass I've just mentioned, all of which are relative, an object's rest mass is intrinsic to it and is thus perspective-independent. In particular, it depends neither on the motion of the object nor on the motion of anyone observing it, which is why it is sometimes called *invariant mass*. It's essential in particle physics not only because the rest mass of a particle is one of its intrinsic properties but also because, as we'll see, all particles of a specific type have exactly the same rest mass. Moreover, it's rest mass that particles obtain from the Higgs field. In other words, the Higgs field is responsible for an intrinsic property of particles, not for a relative one.

Rest mass is the intransigence exhibited by an object when it is pushed by someone who is initially stationary—"at rest," in the dialect of physics—*relative to that object.* If you're moving with respect to an object, nothing stops you from pushing it and seeing how its motion responds, but the intransigence you measure in that way won't be the object's rest mass. Only if the object is initially stationary from your perspective will your measurement of its intransigence be the same as its rest mass.

You are at rest relative to yourself, of course, and so the intransigence of your body, as you experience it when you shift in your chair, begin walking, or stop suddenly, is your rest mass. You're also at rest with respect to most objects you manipulate, such as your kitchen utensils and your cell phone, so your experiences with them involve their rest masses. The same is true of any ball you intend to throw.

The reason rest mass is intrinsic is that it's not open for discussion and argument. Just as different observers may see my workplace as large or small but can all count its four rows of windows, different observers will measure your intransigence differently but cannot dispute your own measurement of yourself. They can all see that you are stationary with respect to yourself, to the cell phone in your hand, and to the shoes on your feet. They can all watch you measure your response to a push, or the response of your phone or footwear. And they can all watch as you record the result of that measurement on a piece of paper. There's nothing perspective-dependent about what you did, learned, and wrote down; they are all beyond debate.

There's a simple reason that Newton didn't realize that mass comes in so many flavors. For an object at rest relative to you, all versions of mass, both intrinsic and relative, are identical. For an object moving at an everyday pace, they differ, but the discrepancies are exceedingly tiny. Not even the planets, moving many miles per second relative to the Earth, easily exhibit these distinctions; for instance, the rest mass, gravitational mass, and other relative versions of mass of Mars, as measured by an observer on Earth, can differ by no more than a few parts per billion. This includes their relativistic mass, mentioned above, which is the intransigence of a moving object when it is pushed forward along the direction of its ongoing motion.

At extreme speeds, however, the differences can be large. If an object zooms past you at 100,000 miles per second, its relativistic mass is 19 percent larger than its rest mass. For a proton accelerated to full speed inside the LHC, the same measurement would reveal a relativistic mass thousands of times greater than its rest mass. But this is from the perspective of someone standing stationary on the Earth. A person traveling along with that proton would instead regard it as stationary, with a relativistic mass no different from its rest mass; see Fig. 11.

Figure 11: (Left) Two identical lightbulbs; the bulb at left, closer to Andrew, seems brighter to him. The reverse is true for Zelda. (Right) Two identical bricks with the same rest mass. From Andrew's perspective (and ours), the lower one moves to the right and has larger intransigence. From Zelda's perspective, the upper one moves to the left and has larger intransigence.

For particle physicists, these distinctions aren't just academic fine points. If someone had accidentally put the wrong version of mass into one of the formulas used during the design of the LHC, the particle accelerator wouldn't have functioned.

I want to consider rest mass from another angle, both because of its centrality in this book and because its definition can be a little hard to follow. In our polymotional cosmos, there's no meaning to saying that "an object is stationary." It would not make sense to say that rest mass is the intransigence of a stationary object. Instead, rest mass is the intransigence of an object *as measured by someone who is stationary relative to the object*. It's a subtle distinction, but it's the difference between meaningless and meaningful.

The fact that the word *relative* appears in the definition of rest mass can be confusing. So let's look at another explicit example that shows why rest mass, even defined in terms of relative motion, is an intrinsic property.

To say that rest mass isn't relative is to say that every one of us agrees on how it should be measured and by whom, so there will never be any dispute about whether one measurement is better than another. The easiest way to find an object's rest mass, we all agree, is to ask what I will call a *privileged observer* to measure it. That's an observer whose motion is the same as that of the object and therefore sees it as stationary.

Let's say you'd like to know the rest mass of my cell phone, but I happen to be passing you in a car. Sure, you could push the cell phone out of my hand and see how it responds; that would tell you its intransigence from your perspective. But that intransigence wouldn't be the phone's rest mass because my phone is moving from your perspective.

To find out my phone's rest mass, you should instead ask me to measure it. Although you view me as moving while I view myself as stationary, we both agree that my motion and that of my phone are identical. That's the key point: we are in agreement that I am privileged and you are not—that I see my phone as stationary and you don't.

So if you or I want to learn the phone's rest mass, we concur that I (or perhaps another passenger in my car, equally stationary relative to the phone) should make the measurement of its intransigence, and you should ask me what the result was. And we agree that it shouldn't be done the other way around!

This is the crucial point: there's no dispute as to who the privileged observers are and aren't (see Fig. 12). That's why we all agree on how much rest mass an object has. Otherwise this would never work; we'd end up arguing.

It may seem inconvenient that whenever you want to know an object's rest mass, you have to hire a privileged observer to carry out the measurement for you. Such an observer might be quite hard to find. Fortunately, there's a way around this. Even if we're not privileged with respect to an object, we can learn its rest mass by

1. Measuring its speed as we see it;
2. Measuring its intransigence as we see it; and
3. Looking up a formula, originally provided by Einstein and checked in countless experiments, that tells us how to infer an object's rest mass from the previous two measurements.[1]

Figure 12: The rest mass of a cell phone in the car is its intransigence as measured by people in the car, relative to whom it is at rest. For a person standing on the sidewalk, the phone's intransigence is larger than its rest mass. All observers agree on these statements.

In other words, rest mass is exactly the same as intransigence for privileged observers, while for all other observers, the object's rest mass is less than its intransigence and must be determined through a more complicated procedure. Fortunately, everyone, privileged and nonprivileged, agrees on the final answer.

By the way, an object needn't be physically close to you for you to measure its rest mass. It's enough that you and the object share parallel steady motion that keeps the separation between your locations constant. While sitting in a chair, you could measure the intransigence of an object lying on a table a mile away, perhaps by firing a strong laser pulse at it. Its motion (as seen by you) in response to that pulse would reflect its intransigence (as seen by you), and in this case, that intransigence would be its rest mass.

There's another way to understand why Galileo's relativity principle requires rest mass to be intrinsic, though in a slightly different sense. The reasoning applies, in fact, to many similar properties of objects.

Without Galileo's principle, there'd be no reason for the basic behavior of the objects around us to be stable from hour to hour, from day to day, from month to month. We take it for granted that chairs that support our weight in the morning can be safely used in the evening and that an oven that can cook a fish in March can cook a similar fish in October. But there's no rule that says a universe has to behave this way.

After all, our motion through empty space changes daily as the Earth spins and seasonally as we orbit the Sun. In a universe where steady motion had intrinsic meaning, the basic behavior of objects around us could vary dramatically as our motion changed over time. We already live with the consequences of daily and seasonal variations in temperature caused by the Earth's relation to the Sun, but imagine if the colors of painted walls, the taste of foods, and the strength of gravity all had similar wide variations simply due to our variable motion through the heavens.

Fortunately, in our universe, this can't be, because of the principle of relativity. For example, imagine that the broiling heat of your oven depended on how quickly your kitchen moved across the universe. Then, by setting your oven to broil and measuring the temperature inside it, you could determine how fast you and your oven were moving through space, violating Galileo's principle. To accord with relativity, then, the temperature created inside an oven set to broil can't depend on how the oven is moving.

The same logic applies to your own intransigence as measured by you, which is nothing other than your rest mass. If your rest mass depended on how you are moving, you could use it to infer your speed across the universe, even in an isolated bubble. The same would be true of the rest mass of any object that travels with you in the bubble. Since relativity asserts that such an inference is impossible, your rest mass and that of the objects that accompany you must be independent of how you are moving, in addition to their independence of how any other observer might be moving relative to you.

The importance of rest mass in science is reflected in its history. The first step was Newton's realization that an object's weight depends on where it is located with respect to other objects, while its mass appeared to Newton to be one of an object's intrinsic properties, independent of its location. Einstein then refined Newton's view, separating rest mass from other types of mass because of its independence of an object's motion as well as its location. When I tell you my rest mass, I need not clarify whether I'm at home, on the Moon, flying across some distant country, or zipping off to Jupiter in a spaceship. For the same reason, the rest mass of an electron is a property that we can be sure of without knowing where the electron is or what it is doing. It's intrinsic to electrons, and that's why it will be so important in this book.

But the true significance of rest mass began to become clearer when scientists discovered and understood *photons*. These are the particles from which light is made. Remarkably, every photon's rest mass is zero.

You've never seen a photon, and yet your eyes have implicit knowledge of their existence. Their particulate nature plays a central role in how you see. Inside your retina are protein molecules called opsins. The sense of vision begins with individual opsin molecules absorbing individual photons; each molecule responds by changing shape, initiating an elaborate process that can lead to an electrical signal being sent down your optic nerve.

However, most forms of light, including radio waves, microwaves, and X-rays, are completely ignored by the human eye, rendering them unseeable and unseen. I'll refer to them collectively as *invisible light*. Importantly, the difference between visible light and invisible light is purely biological: the eye's opsin molecules can absorb the former's photons and not the latter's. As far as the cosmos is concerned, all photons are qualitatively the same. That's why physicists treat all forms of light, visible and invisible, on an equal footing, as we will do throughout this book.

It's well known that light has a characteristic speed, which scientists call c; this is the speed at which each individual photon travels, too. As scientists discovered centuries ago, c is about 186,000 miles per second. That's fast, in a way. Our fastest spaceships don't come anywhere close to that speed. Though my last car was with me for fifteen years, I drove it less than 186,000 miles. At the speed c, you could circle the Earth in a blink of an eye (literally) and travel from my head to my toe in a few billionths of a second.

And yet c is also slow. It takes light more than one second to travel to the Moon, over eight minutes to reach the Sun, and over four years to reach the next-nearest star. If we sent off a robot spaceship at nearly *c* to explore the Milky Way, it could visit only a few dozen nearby stars during our lifetimes.

You and I are small, so we think light runs like a rabbit. But the universe is vast, and from its perspective, light creeps like a turtle.

Light's agonizing crawl is a problem for interplanetary exploration. When the team running the New Horizon mission to Pluto wanted to communicate with their spacecraft, they had an excruciating wait for a reply. Their radio messages, traveling at *c*, took four and a half hours to reach the spacecraft. Whatever answer the craft sent back took another four and a half hours. Just imagine what it's like when your message has a typo and the spacecraft replies, "Could you repeat that, please?" Planetary scientists certainly wish there were a way to speed things up.

A friend of mine who loves languages and words once asked me why physicists refer to the speed of light as *c*. I explained that *c* originates from the Latin word *celeritas*, which means "speed"; our word *accelerate* has the same origin, as does the little-used word *celerity*.

"Oh!" he exclaimed. "I should have thought of that. I had always assumed that *c* referred to light, not speed."

It's just as well that it doesn't, as it turns out that *c* describes not only the speed of light but also the speed of gravitational waves, those ripples in space to which I briefly alluded in the book's opening pages. In fact, any material object must travel at or below this speed. A more accurate name for *c*, then, is the *cosmic speed limit*. This limit isn't a property of light; it's a property of the universe.

Among the various forms of mass, the importance of rest mass becomes clear when you compare objects that have some to those that don't. Roughly speaking, objects that have zero rest mass must travel exactly at the cosmic speed limit, while objects whose rest mass is greater than zero must travel below the limit. On this statement, all observers agree.[2] (There's some fine print that goes with this, most of which we won't need. The only fine print I'll mention now is that my statement applies specifically to objects coursing through empty space. When traveling through a material, such as water or glass, even objects with zero rest mass must move slower than *c*.[3])

Admittedly, there's something strange in this statement. Since speed is perspective-dependent, why would all of us agree on the speed of a flash of light, any more than we'd agree on the speed of anything else? Indeed, there is a hint of inconsistency here, a central one in the history of physics. Its resolution was provided by Einstein himself, as we'll see later in the book.

For the moment, let me ensure that the logic about light's speed in empty space is completely clear. All light—not just visible light of all colors but also all forms of invisible light, from radio waves to ultraviolet light and X-rays and beyond—is made from photons. Photons are particles with zero rest mass. Since all objects with zero rest mass always move at the cosmic speed limit, all forms of light must travel at the speed c.

Conversely, our nonzero rest masses assure that we and all our spacecraft can't ever reach the cosmic speed limit, much less break it. Our speed, relative to any other object with rest mass, must be less than c.

It might be tempting to see a connection between the Higgs phib, which links mass to slowing down, to the fact that objects with rest mass travel more slowly than objects with none. But there's no such connection. Recall the phib's claim: the more the Higgs field slows things down, the more mass they have. (Or perhaps it claims that the more mass they have, the more it slows them down. It can be hard to tell.) Either way, it violates the coasting law and the principle of relativity. By contrast, the true relation between rest mass and c preserves the coasting law: any isolated object with nonzero rest mass will coast and can do so at any speed below c, while an object with zero rest mass coasts at the speed c.

Here's another strange thing. If you have read a variety of books about particles and mass, you will probably have noticed that some say that photons have mass and others say that they don't. It's hard to believe there could be disagreement about something so fundamental in nature. But the origin of the discrepancy is simple: it depends on which version of mass you're asking about.

Photons have zero rest mass, and the fact that they travel at c confirms it. But they have nonzero gravitational mass, which is why they are affected by gravity.

One of Einstein's most famous predictions was that gravity bends the path of light. Not only do astronomers observe this bending, but they often put it to use when looking for dim or dark objects, including black holes, dark matter, and planets around other stars. The gravity of a big black hole, or indeed of any object with large gravitational mass, causes light from the objects behind it to be deflected inward. This distorts our view of these distant objects in much the way that objects' images can be distorted by a lens made of curved glass.[4] The fact that gravity from other objects pulls on light and on its photons implies that photons have gravitational mass.

For an observer who sees an object as stationary, the object's rest mass and gravitational mass are the same. But otherwise its gravitational mass, which is relative, can be larger than its rest mass. Photons are always in motion, so it is perfectly acceptable for them to have both nonzero gravitational mass and zero rest mass.[5]

This distinction is also relevant to another common confusion. Knowing that gravity has something to do with mass and hearing that the Higgs field has something to do with mass, it's easy to jump to the conclusion that the Higgs field and Higgs boson must have something to do with gravity. But in fact, these issues are unrelated. Gravity, associated with gravitational mass, affects all particles, including photons, electrons, and protons, in a universal but perspective-dependent way. In contrast, it is rest mass that's pertinent to the Higgs field, whose influence is perspective-independent yet far from universal. Although the Higgs field provides electrons with the entirety of their rest mass, its impact on protons and neutrons is partial, while it ignores photons entirely.

Historically, the discovery of photons was a conceptual turning point. Earlier scientists had assumed that every object has mass by its very nature. But photons proved this assumption wrong; they are particles without rest mass. That raised a question. Why does *anything* have rest mass?

The problem has a philosophical tinge. A universe in which all the basic objects had zero rest mass, as photons do, would be much simpler than ours. The formulas describing it would be more compact and less complicated. It would be much more in accord with Newton's and Einstein's philosophical view that the workings of the cosmos and the math describing them ought to be beautiful and elegant. If we take this viewpoint seriously, we might wonder why we don't live in such a perfect universe.

Well, before we aspire to such perfection, we should make sure it's really in our best interest.

Physicists are often found imagining universes that don't exist. Does that seem odd? Well, we all do it. We all run scenarios and hypotheticals about the past and the future, and about worlds that differ from our own: *What would have happened if I had gone to a different high school? How will life change if we have a second child? How would society function if no one ever told a lie?* To consider questions about the world as it might be often helps us gain perspective on the world as it is.[6]

In order to understand our own universe better, physicists consider hypothetical universes, asking questions that are often simpler (though stranger) than those we ask in daily life. Translating those questions into math, physicists can often run their scenarios more precisely than humans normally do.

So join me in imagining a universe almost like ours. It's large, expanding, and chilly. It has a cosmic speed limit. But nothing in it has rest mass.

In our universe, light moves at the cosmic speed limit, while everything else we encounter travels more slowly. In this imagined universe, every object moves at the limit. No one stops to look at the stars. No one stops to smell the roses. No one stops to think. No one stops. No one even slows down.

Worse, there are no stars, or roses, or thoughts. A star must form; a rose must grow; a thought needs a thinker. But how can a ball of gas collapse to create a star, or a seed germinate, or a brain be wired together if all the world's ingredients are forever moving at the limit? What sandcastles can you build from sand that refuses to stay in place, instantly flying away as though blown by a hurricane?

This restless universe is much simpler than our own. It looks the same everywhere. It has the same appearance through a strong microscope as it does through a powerful telescope: smooth, perfect, unblemished, like a white wall painted by machine. It is a most elegant universe. Elegant, and lifeless.

Fortunately, the universe we know isn't elegant. It is violent, chaotic, diverse, full of structure and complexity. Why? What crucial ingredient does our universe possess that its restless, lifeless cousin lacks?

What makes our universe inelegant but more vibrant than a restless universe are its objects, among them electrons, protons, and neutrons, with rest mass. These particles can—no, must, by law—travel below the cosmic speed limit. Their ability to slow down is what allows them to form atoms. Since atoms must move below the speed limit, giant clouds of them can be gathered by gravity and can collapse under its pull to form stars and their planets. On those planets, the atoms can bind together, forming molecules of water, sugar, minerals, and proteins. These can be assembled into DNA, into cells, into plants and animals, into rocks and roses, and into brains with thoughts.

It is rest mass that makes it all possible: the atoms, stars, roses, books, and brains of our living universe.

This fable doesn't explain why objects in our universe have rest mass. Addressing that issue, at least in part, is a goal of later chapters. But it does tell us that in a universe where nothing had rest mass, there'd be no one to ask why.

My critique of this imaginary perfect cosmos might seem a gentle gibe at my friend and colleague Brian Greene and his famous book *The Elegant Universe*. But really, I'm poking harmless fun at a whole worldview, one that goes back all the way to Einstein, even to Newton. Einstein believed that the equations describing nature should be simple and elegant. Sometimes this viewpoint served him well and sometimes it didn't, but it has been absorbed into the cultural aesthetics of many theoretical physicists, Greene among them.

Though elegance might be fine for formulas that describe the principles underlying a universe, it's not so good for the universe itself. Luckily, we can have it both ways. As a number of human games illustrate, a few straightforward rules can still lead to a fascinating, complex outcome. An astonishingly intricate, disorderly cosmos can emerge from simple natural laws that humans would regard as aesthetically appealing.

Many physicists do expect that the laws of nature will turn out to be beautiful and simple. But let's be clear: this is a hope, a theoretical speculation. If you look carefully at the laws we know so far, there's not much evidence that they're intrinsically elegant.

One of the striking things about the formulas we use in modern physics is that they're not that complicated, but they're not so simple, either. To the extent that they are simple, it's not because of a principle. It's a result of limited knowledge. Partial information can make something appear more ideal and refined than it really is. The Earth, for example, appears to be a perfectly round blue-white marble when viewed from afar. Its apparent perfection reflects one's inability, at that distance, to recognize its seas, continents, mountain ranges, clouds, and deserts, not to mention its slight equatorial bulge.

This viewpoint on our equations is due to legendary physicists of the 1960s and 1970s, among them Leo Kadanoff and Kenneth Wilson.[7] It warns us that we are not necessarily better off than the natural philosophers who long preceded us, many of whom imagined the Sun and Moon as perfect spheres (until Galileo and his contemporaries observed the Sun's spots and the Moon's craters), planetary orbits as perfect circles (until Kepler realized they were ellipses), and water as a continuous substance (until Einstein and his contemporaries established the existence of water molecules). We simply do not know yet whether the underlying laws of nature are elegant or not.

As you will see later in this book, the Higgs field exhibits the most inelegant of the known laws governing fields and particles. There's an amusing tendency for those who tout beauty to ignore this, as though it were an inconvenient family member, and to focus instead on Einstein's elegant theory of gravity. Yet even that theory has its issues.[8]

The idea that nature's rules should be elegant is a bias. Maybe it's correct, but in doing science, we must be careful not to project our biases onto the universe. It's okay for individual scientists to have biases, as long as they are aware of them and as long as diversity among our scientists guarantees a diversity of biases. If not, we'll head off together in the wrong direction, trying to explain why the Sun's a perfect sphere without noticing that it isn't. No matter what our personal inclinations, we must always let the universe speak for itself; that's why experiments and observations, not theoretical reasoning, are the foundation of modern science.

6

Worlds Within Worlds
The Structure of Material

Many people remember their first chemistry class better and more fondly than any physics class they may have taken. Chemistry is more familiar: ingredients, instructions, step-by-step activities, interesting outcomes with weird smells. It's much like cooking. Indeed, to a large extent, cooking *is* chemistry.[1]

Also, physics class just doesn't stick as well. It's abstract, counterintuitive, complicated, and not as much fun as playing with chemicals. I don't take this personally. I loved eighth grade chemistry before physics swept me away in ninth.

Since simple chemistry often precedes physics in the learning sequence, it's where mass and various other scientific ideas are first taught. The problem here is that certain statements made in chemistry books, though (almost) true in daily life and in a chemistry lab, fail badly in other contexts. These statements must be unlearned if we are to understand the cosmos.

We've already encountered one lesson to unlearn: that mass is the quantity of matter. Related to this lesson is another: in chemistry class, we are taught that mass is "conserved"—a false friend that means "preserved" or "unchanged" in English. Mass, says the chemistry textbook, is never lost or gained; however much you start with is what you end with.

This amazing fact about chemical reactions was demonstrated experimentally, within a century after Newton, by Antoine Lavoisier, one of the founders of modern chemistry. It's still taught to students, just as Newton's laws are, without warning them that it's not entirely true.[2]

The origin of this lesson of chemistry class is that in all chemical processes, atoms are rearranged into new patterns but are never created, destroyed, or profoundly altered. Indeed, this is the definition of a chemical reaction. That's in contrast to a nuclear reaction, such as nuclear fission and fusion, in which atoms can be changed from one type (or *element*) to another. In other words, chemistry involves processes in which the atoms themselves are preserved (i.e., conserved, in physics dialect). Because each type of atom has a fixed, definite rest mass, the preservation of the atoms from the beginning to the end of a process assures that the total rest mass of everything involved can't change, either—at least not by an amount that matters in chemistry class.

In most ordinary experiences, too, rest mass appears to be conserved. When you crinkle up a piece of paper, its rest mass stays the same. When a puddle of water freezes, it expands, but its rest mass doesn't change. When you add salt to water and the salt dissolves, the rest mass of the salt water is equal to the rest mass of the fresh water you started with plus the rest mass of the salt you added. And so on.

In daily life, if an object gains rest mass, it's because extra rest mass arrived from somewhere else, and if it loses some, it's because some rest mass has gone elsewhere. The total amount of rest mass in the universe hasn't been altered.

But in any physical process in which atoms can be destroyed, created, or profoundly changed, or in which atoms play no role, rest mass need not be conserved. It may increase or decrease; the accounts will not balance.

This isn't just of academic interest. If rest mass couldn't decrease inside stars, their nuclear furnaces wouldn't ignite. The Sun wouldn't shine, and Earth would be a frozen rock.[3] Even at a more mundane level, decreasing rest mass makes possible various medical procedures and diagnostics, including certain cancer treatments and PET scans. And the LHC couldn't make Higgs bosons if rest mass couldn't increase.

Chemistry class, attributing the rest mass of an ordinary object to its atoms, takes the rest masses of their subatomic constituents—electrons, protons, and neutrons—for granted. That's why, to understand mass, we must go beyond chemistry to the subatomic realm. This will take us to Einstein and to his most famous formula.

Toward that end, let's take a brief tour of the structure of ordinary material, descending a ladder from human size down to the subatomic. A secondary purpose of the tour is to give you a sense of scale, which will be useful both in this book and, I hope, beyond. But before we even start, I want to equip you with a powerful tool, one of the most important and least appreciated among those in a physicist's toolbox.

Scientists have a reputation as precise thinkers, and physicists are often imagined as the most obsessively, tediously precise. Yet one of the most important skills young physicists acquire is a sense of when and how to be *imprecise.*

Sure, I've always had a flair for math. A friendly joker in my sixth grade class used to refer to me as "the Computer Who Wore Boots." As a physics student, I learned how to do complex and detailed calculations. But what makes a physicist more powerful than any current computer is that we have been taught when to be precise... and when not to be.

Rational thought is essential to science, but one also needs an intuitive sense of how meticulous or vague to be as one engages with nature. Without this intuition, the natural world cannot be grasped; it's far too complicated, with too many details to keep track of.

This skill might sound exotic, but you already have something very much like it. If you were asked when you were born, how would you reply? You might just say, "Nineteen seventy-one." Or you might give a date: "February 7, 1971." You might even include the hour, perhaps noting the time zone. You would, of course, choose your answer based on the context of the question.

If a stranger asked where you live, context would determine whether you would answer with your country, your city, your street, or even your complete mailing address. Rarely would you say, "Earth," or give latitude and longitude coordinates, though you can probably imagine improbable situations in which such responses would be appropriate.

Breaking these rules easily draws a laugh. Kids, bless them, do it all the time out of naivete. It's a ready joke in movies: a scientist's age given to the nearest minute. Or the person buying a gift abroad who, when asked for a shipping address, responds, "Oh, everybody knows me back home. Just put my name on it and write 'London.'"

It's no surprise, then, that students find it hilarious when I stand next to a meter-long ruler and describe myself as "roughly one meter tall." A five-year-old might be a meter tall, but a typical adult's height is much closer to two meters than to one. Why would I begin a physics class by undermining my own authority, confidently claiming something that is obviously absurd?

In fact, the statement that I am about one meter tall is neither absurd nor false, though making sense of it requires an understanding of the physicist's rule of precision. What draws my students' laughter is that I seem to have violated the rule, but what I'm teaching them at that moment is that I haven't.

In a moment we will find ourselves zooming back and forth between atoms, galaxies, subatomic particles, humans, and the universe. We will be considering billions of years and trillionths of trillionths of seconds. The difference between two meters and one meter will be an irrelevant and distracting detail, akin to the excess baggage found in answering "How old are you?" to the nearest minute.

The rule of precision is simple: be as precise as necessary to answer the question, understand the concept, or otherwise serve the purpose, but be no more precise than that. It's a familiar principle, known to everyone. Being a physicist requires extending it to the unfamiliar settings of the universe.

The rule of precision plays a central role in the sense of scale, an intuitive sense that can help us remember which microscopic things are larger than others and, very roughly, by how much. You can think of it as a mental cheat sheet that helps us keep track of how the universe works. It's an essential tool in a physicist's toolbox, and it requires judicious imprecision; that's the trick that helps make the relationships between important objects easier to remember.

Our tour of the structure of ordinary material will take us down a ladder of ever-shrinking sizes. Each rung in the ladder will have us jumping down a step of a hundred thousand or so.[4] But here I'm expressing myself imprecisely.

When I say that "I'm in my fifties," the wording makes clear that it's an approximate statement. Yet sometimes, by referring to a person's age as "sixty," we may implicitly mean "sixtyish" without saying so. That's going to be our situation here. On our tour, I'll often use numbers like 100,000 (or other numbers with a 1 and a bunch of zeroes), but I will never mean such simple numbers to be exact.[5]

I realize, though, that the number 100,000 is larger than most of us can easily imagine. From teaching, I have found some effective ways to conceptualize 100,000 in an approximate way. It's roughly the number of people in a large town or small city; the number of seconds in a day or the number of times your heart beats each day; the number of steps taken on a walk of roughly 50 miles (80 km); the number of minutes in a summer; the number of human body lengths traveled while driving two hours on a highway; and the number of words in this book. For me personally, the last one's my favorite, with the previous one a close second. But you should find the comparison that speaks most vividly to you.

Scientists have a useful shorthand that makes big numbers easier to describe, and I'll use it occasionally. The idea is that since 100,000 has five zeroes following a one, we can write it as 10^5. (You can either just think of this as an abbreviation, a way of saving you the trouble of counting all the zeroes in a big number, or if you like math, think of it as a true exponent, five powers of ten multiplied together.[6])

6.1 From the Human to the Cell

Armed with judicious imprecision, let's begin by imagining your body is abruptly expanded 100,000 times in length, width, and height. Your two feet could stretch the distance between cities separated by an hour or two by car (New York to Philadelphia, London to Birmingham, or Kobe to Kyoto). Your head would be at the very edge of the Earth's atmosphere.

The rest of us, bewildered, would have a hard time seeing you as a whole. But human beings are best understood not as individuals but rather as societies of living creatures, called cells, that live and work together, carrying out different functions that collectively benefit the whole human organism. Since the length of a typical human cell is roughly 100,000 times smaller than the height of a human, your cells, inflated along with the rest of you, would be easy for us to see.

The quantity of cells in a human body differs from person to person and from day to day. Moreover, when counting our cells, we must decide whether to include the many bacteria living inside us. But roughly speaking, our cells number 10^{13} to 10^{14}—between ten and a hundred trillion.

Even this imprecise estimate teaches us an important lesson: a human's cells are roughly a thousand times more numerous than the Earth's entire human population (roughly ten billion, or 10^{10}). As a microbiologist I know put it, "Human societies are far less populous than human bodies are cellulous."

You might wonder, considering how chaotic and counterproductive human societies are, how it is possible for a human body to remain organized and functional with such an immense number of cells. How do all those creatures manage to avoid working at cross-purposes? What a fantastic set of questions biologists get to contemplate, far beyond what methods in physics can handle.

6.2 From the Cell to the Atom

Let's now take another step. What's smaller than a cell by 100,000 in length, width, and height? Here, leaving the living world behind, we find the basic building blocks of ordinary material: atoms such as hydrogen, iron, and gold, along with small molecules made from a few atoms such as water, carbon dioxide, and methane (the natural gas that heats many homes).

Cells can be photographed using a microscope. That won't work for atoms, so tiny that visible light can't bounce off them effectively.[7] Still, atoms can be imaged in several ways. For instance, if a beam of electrons is pointed at a thin surface, one can measure the impact of the individual atoms on the beam as it passes through the surface. The atoms' locations and sizes can be inferred, and that information can then be turned into an image, an example of which is shown in Fig. 13. Each "ball" in the image is an atom, about $100,000 = 10^5$ times smaller in width than a cell and about $10,000,000,000 = 10^{10}$ times smaller than a human is tall.

Figure 13: An image of atoms of different elements in a crystal made by a transmission electron microscope. Rather than detecting the reflection of visible light, this microscope detects how the material blocks a narrow electron beam. Image cropped and rotated.

Scientists deduced the size of atoms, using clever indirect reasoning, long before such microscopes existed. The issue was on Einstein's mind in 1905, both in his doctoral thesis and in one of his other famous scientific papers of that year. His work, and that of other scientists of that era, convinced the last skeptics that atoms really do exist.

Once scientists knew the diameter of a typical atom, they could estimate the number of atoms in the typical human body. That number?

$$10,000,000,000,000,000,000,000,000,000$$

That's 10^{28}. I invite you to count all twenty-eight zeroes and make sure I didn't miss one.

Now, I don't know about you, but this staggering number gives me vertigo. Think about it. It means that the atoms in a human body are far, far more numerous than the sand grains on all the beaches of the Earth; than the leaves and needles on all the trees of the Earth's forests; than the number of seconds since the turmoil of the Big Bang; than the heartbeats of all the people who have ever lived upon our planet; than even the stars—not just the ones you can see in a dark sky, not just the hundreds of billions in our Milky Way galaxy—far more numerous than *all* the stars in *all* the galaxies across the entire visible universe.

I doubt I can convince you that a human brain can make sense of a number so ridiculously big. My own brain cannot. It's astonishing that such a prodigious crowd of atoms can remain organized as a living creature.

This came up in conversation with an engineer I know. "In a sense," she mused, "I've known all this since high school chemistry, when I first encountered Avogadro's number.[8] But I never really thought about it. It was just a number that went into equations. I never saw it in the mirror."

I don't think she's atypical. Even though this number stares back at me every time I look down at my own hands, it still feels terribly abstract, despite my decades as a scientist. And yet there's nothing abstract about it. It represents instead something fundamental about concrete reality, a central fact of nature: the recipe for a single human body requires far more atoms than all humans put together could ever hope to count.

Our brains aren't designed to deal with this knowledge. Physicists learn to use math to help us cope with the tiny and the immense, allowing us to draw accurate conclusions about the cosmos while remaining sane. But visualizing a number with twenty-eight zeroes in it?

I try, though. Every now and then, I come up with a new approach. Here's one: How many human beings, packed tightly like sardines, could fit inside the Sun?

6.3 From the Atom to the Atomic Nucleus

The idea that material things might be made from atoms dates back at least to the ancient Greeks. It may have originated with the protoscientist Democritus, or perhaps his teacher Leucippus. The word *atom* was chosen because *atomos* means "uncuttable" or "indivisible." Scientists started finding evidence for the existence of atoms in the early nineteenth century.

But as happens so often in science, the term *atom* is a misnomer. Thank goodness atoms are, in fact, divisible! We wouldn't have chemistry and biology otherwise.

Let's peer into an atom. The best microscopes don't help; they show us atoms that look like little balls. This is a mere illusion. If you took a photograph of a hovering hummingbird with a camera whose shutter speed was a full second long, the image would be just a blur, with barely any sense of the shape of the bird and no sign of its wings. Such is our problem here. The microscope operates far too slowly, so its picture of an atom is too blurry to show us its internal structure.

In fact, there are fundamental limitations on how quickly an image can be made. At the atomic scale, we and our machines lose the capacity to see or feel directly; no microscope or camera could ever do the job. Instead, we must take increasingly indirect and subtle approaches in our exploration of nature. Our conclusions are no less confident, but there's no picture to say a thousand words. That's why, after telling you what we know, I'll give you a little sense of how we came to know it.

You've likely seen drawings of atoms that look like Fig. 14, with electrons on the outside orbiting an atomic nucleus made of protons and neutrons.[9] This symbolic image encodes some correct facts about atoms, but it is intended only as a cartoon. It's much less accurate than a stick-figure drawing of a person.

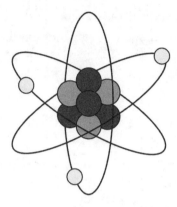

Figure 14: The universal atomic cartoon, whose electrons orbit a nucleus made of protons and neutrons, drawn in different shading to distinguish them. A real atom is very different; the nucleus is tiny compared to the atom, and the electrons even smaller.

The problem is one of proportion, so wildly distorted that the atomic cartoon barely resembles an atom at all. For starters, the atom's nucleus is drawn far too large; it should be 100,000 times smaller, in length, width, and height, than the atom itself. The electrons, in turn, are even smaller than that!

Consequently, the vast majority of an atom—and of everything made from atoms, including you and me—is *empty space.*

Imagine that we took an atom and inflated it, along with its electrons and nucleus, so that it became the size of a typical high school or college classroom. The atom's nucleus would then be the size of a tiny sand grain. The electrons, whizzing around at incredible speeds, would be far too small to see. Beyond this, the room would be entirely void.

In short, every atom is as empty as a lecture hall with all its furniture and its air removed, leaving it vacant except for a floating grain of sand.

We're made mostly of nothing. So is everything we eat, everything we touch, every mountain we climb, our entire planet. Almost—but not quite—nothing.

Upon this revelation, a friend of mine began to protest vehemently. "That's absurd, Matt!" he objected. "It makes no sense. If I'm mostly empty, and my chair is mostly empty, why am I not falling through my chair?"

I agreed with him that it violates common sense. Material things don't feel or look empty to me, either. Our blood stays inside our veins and our air inside our lungs. The food we eat seems substantial. We don't sink into the Earth.

Nevertheless, what we experience as ordinary existence is an illusion created for us by our brains. It is based on the interaction of our senses with the physical world, not on the physical world directly. The illusion has to help us survive in the real world, so it had better have something to do with reality; it needs to inform us when there's a fruit tree or a hungry tiger nearby. But beyond that, there's no reason that it needs to be scientifically accurate. The reason that many objects in the world seem opaque and impenetrable is this: for most practical purposes in human life, they act as though they are. But while evolution focuses on what is practical, physics is about what is real, and there's no reason they should agree.

At the end of this book, I'll explain why we don't sink through the ground or through chairs—why atoms, though empty, are nevertheless impenetrable *to other atoms*. At an intermediate stage, we'll see that atoms are a little less empty than I've implied here. But what I've just told you is largely true, and here's how we came to know it.

That the nucleus is tiny and the atom empty was learned in 1911, when Ernest Rutherford, a New Zealand–born British scientist already graced with a Nobel Prize for his work on radioactivity, correctly interpreted a surprising experiment done at his laboratory by his assistant Hans Geiger[10] and his student Ernest Marsden. The two younger scientists aimed a beam of fast-moving particles at a thin sheet of gold atoms. Most of the particles went right through, as though there were nothing there for them to hit. But a very small fraction bounced back, proving to Rutherford that there was something tiny but hard in the heart of an atom.

Another early clue was the discovery of X-rays, whose photons mostly pass through a human body and even through thin walls. If atoms weren't empty, that wouldn't happen.

Stars provide another spectacular demonstration. A large and dying star can implode, collapsing inward under the force of its own gravity and crushing itself down to form a dense ball of neutrons called a *neutron star*. Though it has the mass of an ordinary star, a neutron star is much smaller than the Earth. This shows that ordinary material can be dramatically compressed, serving as additional evidence of atoms' remarkable emptiness.

6.4 From the Atomic Nucleus to the Edge of Knowledge

A nucleus can be envisioned as a rather tightly packed clump of protons and neutrons, each of which is about 100,000 times smaller than an atom. In that sense, the atomic cartoon's sketch of a nucleus, though drawn about 100,000 times too wide, isn't wildly off in its shape (Fig. 15).

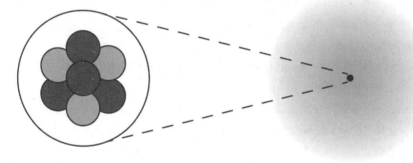

Figure 15: A better cartoon of an atom, with a tiny nucleus of protons and neutrons surrounded by a cloud of even tinier electrons. It is still not to scale.

Protons were discovered in the 1910s and neutrons in the 1930s. But only in the late 1950s did scientists learn that these objects, too, have a measurable size and are not elementary. They then asked what's inside them. Are they hard balls? Are they soft, mushy blobs? Are they made from even smaller objects?

In the late 1960s, using a method similar to the one that led to Rutherford's breakthrough, physicists obtained an answer. By firing electrons at protons and measuring what happens to them, scientists learned that protons and neutrons are far more complex than atoms. Each one is made from an army of particles that rush around at or near the cosmic speed limit, pulling vigorously on one another. This type of pull, as fundamental in nature as gravitational or electric forces, is called the *strong nuclear force*.

My description of a mob of swift particles inside a proton may very well contradict what you've read elsewhere. It's stated in many websites and books that a proton is made merely of two up quarks and one down quark and nothing more. I've illustrated this in Fig. 16 at left, representing the up and down quarks as u and d.

(Despite their names, there's nothing actually up-ish or down-ish about these quarks. Up quarks are one type of particle, down quarks another; that's all. The names originate from a series of historical accidents, just like many other words in physics dialect, and they could just as well have been called Fred quarks and Alice quarks. Similarly, there's nothing strange about the "strange quark," a third type. The names are whimsical, but they've stuck.)

This simple three-quark picture, an antiquated idea from the 1960s, is the proton phib. It's still used today to keep explanations short and simple, at the cost of accuracy. There's a grain of truth in it, as we'll see in a moment. But we try to avoid phibs here.

By the early 1970s, scientists had realized that quarks don't just drift around the proton—they bounce around violently within it. Moreover, they are accompanied by another type of particle, which was named the *gluon*, from the word *glue*. As we'll see much later, these particles are loosely associated with the strong nuclear force, which, poetically speaking, "glues" the proton together. Gluons, which I'll represent with a g, are abundant inside the proton, as shown at the center of Fig. 16.

That's not quite all. The proton also contains some pairs of quarks and anti-quarks. Before I say more about them, I should first explain what anti-quarks are.[11]

It turns out that in any universe governed by Einstein's version of relativity, for every type of particle there must be another type that serves as its antiparticle. The up quark's antiparticle is the up anti-quark. Consequently,

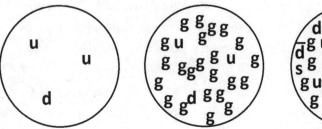

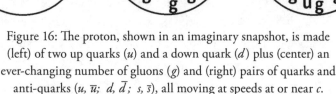

Figure 16: The proton, shown in an imaginary snapshot, is made
(left) of two up quarks (*u*) and a down quark (*d*) plus (center) an
ever-changing number of gluons (*g*) and (right) pairs of quarks and
anti-quarks (*u, ū; d, d̄; s, s̄*), all moving at speeds at or near *c*.

the up anti-quark's antiparticle is the up quark. Similarly, the down quark's
antiparticle is the down anti-quark, and vice versa.[12]

Importantly, this is a relation between *types* of particles, not between individual particles. Einstein's version of relativity implies that if electrons exist in a universe, then their antiparticles, called positrons, must exist there,
too. The reverse is also true. But this is not to say that every individual electron has a positron partner. In fact, our universe has many more electrons
than positrons. Similarly, there are more quarks than anti-quarks. We don't
know all the reasons behind this so-called matter/antimatter asymmetry; its
origin, though not a focus of this book, is a long-standing puzzle.

In some cases, a type of particle is its own antiparticle. For example, the
photon is its own antiparticle; there is no separate anti-photon. The same is
true of gluons, Higgs bosons, and a few others. (The origin of this pattern,
though simple, isn't relevant here; I'll say more about it in a later chapter.)

Now, back to the proton. In addition to its three famous quarks and its
gluons, it contains pairs of up quarks and up anti-quarks. I'll indicate them
by adding an equal number of *u* and *ū* letters to my depiction. Similarly,
there are pairs of down quarks and down anti-quarks (equal numbers of *d*
and *d̄*) and occasional pairs of strange quarks and strange anti-quarks (equal
numbers of *s* and *s̄*).

This mess of gluons, quarks, and anti-quarks, as rendered at the right of Fig. 16, is about as illuminating a cartoon of a proton as I can manage. Yet it's misleading in three ways. First, although the picture looks crowded, it's simply because the letters representing the particles have to be large enough to be read. Compared to the corresponding letter, each quark and gluon is tiny. For this reason, the proton, like an atom, is relatively empty. Second, my depiction is at best a sort of imaginary snapshot. The particles in the proton are rattling around at or near the cosmic speed limit. In contrast to an atom's elegant, highly organized structure, a proton is more of a madhouse. Third, as we'll touch on later, quantum physics alters our basic conceptions of quarks and gluons and forces us to rethink the picture; this is true for electrons and our cartoons of atoms, too.

Despite the fact that a proton is far more complicated than the proton phib would suggest, there's a remnant of the phib inside the truth. If you were to collect all the up quarks and up anti-quarks and pair them, one anti-quark for each quark, you'd find two *extra* up quarks left over. Similarly, you'd find one extra down quark, while the strange quarks and strange anti-quarks would pair up perfectly.

One may say, then, that 1970s scientists replaced the three-quark picture of the 1960s with a three-extra-quark picture. In this account, two up quarks and one down quark are engulfed in a bath of gluons and quark/anti-quark pairs. As these particles fly around at high speed and collide with one another, the number of gluons and quark/anti-quark pairs is always in flux. Sometimes two gluons will collide and turn into a quark and anti-quark of the same type, and vice versa; these collisions can also produce or absorb other gluons. Yet one thing remains constant amid all these transformations: within a throng of other particles, there are always three extra quarks, two of them up-quarks and one a down-quark. In the sense of physics dialect, the number of quarks minus the number of anti-quarks is conserved. And so, once an object with three extra quarks has somehow been formed, the number of its extra quarks never grows or shrinks and forever remains three, unless and until the object is destroyed.[13]

Considering all this complexity, you can understand why the proton phib has survived; it's a quick sound bite, whereas the truth requires several pages.[14] But I don't think it's wise for physicists to misrepresent the most common particles in our bodies and our surroundings out of concern that they're too complicated for nonscientists to comprehend. Moreover, the complications are relevant in this book. To appreciate where our rest mass comes from, and to be clear on the Higgs field's role in nature, we need to picture a proton more accurately.

The neutron is nearly the proton's twin. To create a neutron, simply replace one of a proton's up quarks with a down quark. This explains the neutron's phib, according to which it consists of one up and two down quarks. In reality, its interior, like that of a proton, includes a bath of other particles.[15]

Before moving on, let me address an issue that might puzzle some readers. It's widely claimed that when antimatter meets matter, a powerful explosion results in which both are destroyed. By all definitions of *matter*, protons and neutrons are examples of matter, so you would naturally guess that quarks are matter and anti-quarks are antimatter. How, then, can there safely be both quarks and anti-quarks in our bodies? Why don't our protons and neutrons spontaneously detonate?

Here we face the antimatter phib, both an oversimplification and an overstatement about how particles behave. It's true that if a large object made of anti-atoms, themselves made of anti-protons and anti-neutrons surrounded by positrons, were to meet a large object made of ordinary atoms, there would be a big explosion. But that's a specific statement about significant quantities of atoms and anti-atoms, not a statement about antistuff in general. When a single particle of one type meets a single particle of its antiparticle type, there are many things that can happen. For example, a collision may convert an up quark and an up anti-quark into two or three gluons, or into a down quark and a down anti-quark, or even leave them unchanged. This is not the full list of options, either. Each of these processes can also happen in reverse. Such collisions and transformations are happening continuously and stably inside protons and neutrons. They are completely benign, cause no damage, and always conserve the number of extra quarks.[16]

6.5 End of the Tour

As of 2023, we can go no further. By using the LHC as a quasi-microscope,[17] we have learned that electrons, quarks (and anti-quarks), gluons, and various friends of theirs are at least 100,000 times smaller than an atomic nucleus. For all we know, they may be much, much smaller than that; they may even be points of infinitely small size and may not be made from anything else. Since they might be among the most fundamental objects in the universe, we refer to them as elementary particles—elementary in the sense of "elemental."

When I explained this to a few friends over dinner, one of them, a lawyer, sat up straight. "Wait a bloody minute!" he exclaimed. "If you don't actually know they're elementary, what right do you have to call them that?"

"Oh, we don't claim to have any such right," I told him. "We physicists know perfectly well that 'elementary particle' is misleading shorthand."

Scientists once thought atoms were elementary—which is why we're still teaching the Periodic Table of the Elements in chemistry classes. And not so long ago, scientists imagined that protons and neutrons might be elementary. All we can really say, if we want to be accurate, is that electrons and quarks are up-to-now-apparently-elementary particles.

But when I proposed that ungainly phrase as an alternative, everyone around the table grimaced. "Exactly," I laughed. "Such a name would be much more accurate, but it's long and clumsy, and nobody would use it."

It's true that shorthand in our dialect often obscures our meaning. But it's much the same in any professional dialect: abbreviations are used to summarize complicated ideas so as to streamline discussion. This makes life easier for the experts, at the cost of making their conversations more difficult for others to follow.

Our tour now comes to an end. By leaps of 100,000, we have stepped down from the human to the cellular to the atomic to the nuclear. From there we have descended a final step to the edge of knowledge, where we find a frontier populated by objects whose size cannot yet be measured with modern technology, the "elementary" particles out of which we are made. Along the way, we have contemplated the hordes of particles in our bodies, the near empti-ness of atoms, and the inner complexity of protons and neutrons.

Whether such tours will someday proceed further—whether the up-to-now-apparently-elementary particles are themselves made from something even smaller—is a question answerable only by future experiments, to be carried out by our descendants. For now, this is what we know.

7

What Mass Is (and Isn't)

Suppose I handed you a bag of white rice and asked you to figure out its rest mass. If you had no way to measure its weight, what might you do? Based on the principle that ten glass marbles have ten times the mass of one marble, you could roughly ascertain the bag's mass using judicious imprecision. You could first estimate how many grains of rice are in the bag. Then you could look up the rest mass of a typical grain of white rice. Multiplying these two quantities together would give you an estimate of the whole bag's rest mass.

You could similarly estimate your own rest mass, with your protons and neutrons playing the role of the grains of rice and your body playing the role of the bag. Almost exactly, your body's rest mass is just the rest mass of all its atoms added together. Each atom's rest mass is almost exactly the rest mass of its nucleus, because electrons have much smaller rest masses than do protons and neutrons. And the rest mass of each atomic nucleus is almost equal, to within 1 percent, to the sum of the rest masses of all its protons and the rest masses of all its neutrons.

Crucially, all protons are identical, as are all neutrons. (We will learn why before this book is over.) On top of that, the rest mass of a proton is almost the same as the rest mass of a neutron, to better than 1 percent. So your rest mass, to an accuracy of better than 1 percent, is simply the total number of protons and neutrons in your body multiplied by the rest mass of one proton (or one neutron).

The mass of an ordinary object, then, can be said to be a measure of the quantity of protons and neutrons it contains. You could call that "the quantity of matter" or "the amount of matter" if you like; it's a matter of definition. And this is why Newton's belief that mass is the quantity (or amount) of matter survives in twenty-first century dictionaries and chemistry classes.

But if we want to understand rest mass, this insight doesn't really help us. It just forces us to ask a new question: Where does the rest mass of a proton or a neutron come from? Besides, elementary particles such as electrons, quarks, and Higgs bosons don't have protons and neutrons inside them. Their rest masses will require some other type of explanation.

Our problems aren't limited to particles, either; similar issues arise for black holes. (Sorry for the sudden swerve from the subatomic to wacky space things, but there's a point to it.) One could make a big black hole entirely out of electrons and positrons, or even out of photons, in which case it could have a large mass and large size without eating a single proton or neutron. So even macroscopic objects, unfamiliar ones to be sure, can have a rest mass unrelated to the numbers of neutrons and protons inside them—and therefore, nothing to do with the quantity of their matter by any possible definition of the term.

For physicists of the present, no notion of *matter* or *stuff* would allow us to retain Newton's definition of mass. We need a different approach.

7.1 Two Birds with One Stone

Modern physics is founded on two fundamental relations. Both were guessed, and expressed as formulas in two unrelated scientific articles, by a single person back in 1905. The author was twenty-six years old, busy finishing his doctoral thesis at the time. Part of that is typical—I also got my doctorate in physics at the age of twenty-six—but on the scale of what this fellow achieved, I would describe my scientific accomplishments as invisible. And I wasn't supporting a family by working at a patent office, either.

Yes, I'm referring to Albert Einstein. Most other students pursuing doctoral studies in physics made ends meet by teaching or doing research. But Einstein, while studying physics as an undergraduate at a university in Zurich, managed to annoy an influential professor. Misreading Einstein's ability and peeved by his youthful arrogance, this professor essentially blacklisted him, leaving him unable to follow the usual route to a doctorate. That's why, while living in Bern and preparing to submit his thesis to a second university in Zurich, young Einstein had to find an unusual source of income. Hence the job at the patent office in Bern. Fortunately for physics, Einstein didn't let this stop him. This was thanks both to his exceptional intellect and to several friends, also physics and math doctoral students, who provided him with a community as well as access to physics journals in the local university library.

The two revolutionary formulas of 1905 each express an unexpected relationship, linking two quantities that people once thought had nothing to do with each other. The first is the one that everyone has heard of: Einstein's relativity formula, $E = m[c^2]$. I've added the brackets to emphasize that c^2 plays a minor role. What's important in the formula is that it tells you that E and m are related. The cosmic speed limit c is unchanging, so if you want to switch between E and m or vice versa, the quantity c^2 acts merely as a conversion factor, similar to the one needed to switch from kilometers to miles.[1]

The second relation, less widely known, is the quantum formula, which reads $E = f[h]$. It holds the secrets to atoms, to light, and to all of quantum physics, including modern electronics and computers. It was first written down in 1900 by Max Planck, a generation older than Einstein and nearly as influential. The quantity in brackets, known as Planck's constant, is again something that never changes, and so the formula relates E and f, with $[h]$ serving as a conversion factor.

Thus, we have two formulas relating E to something else. Even if you know what E and m and f stand for, watch out! We shouldn't instantly assume that E denotes the same thing in both formulas—it does, and yet it doesn't—and we'll soon see how important it is to be clear and specific about what E, m, and f represent. We will even need to be careful about what "="

signifies! Otherwise the meaning of the formulas will be obscured, and the universe will remain out of focus.

All famous things and people are shrouded in a fog of misinformation, and the historical haze surrounding Einstein and his formulas is particularly dense.[2] When I was a child, among the first things I learned about Einstein were that he failed eighth grade math, that he was an untrained and isolated patent clerk who single-handedly revolutionized physics, and that he played a central role in the development of nuclear bombs. Every one of these lessons is a myth.[3]

I clearly remember one of my fourth grade teachers telling our class a story of how $E = m[c^2]$ contributed to the creation of nuclear weapons. He explained that when Einstein first wrote it on the blackboard, people in the room, immediately grasping its terrible implications, burst into tears. But this is impossible. At the time, Einstein was an unknown student, and the atomic nucleus hadn't been discovered yet. Although radioactivity had recently been observed and studied, notably by Henri Becquerel and by Marie Skłodowska Curie and her husband, Pierre Curie (all winners of the 1903 Nobel Prize in Physics), knowledge at the time suggested a steady source of energy, not an explosive. I have long wondered how my teacher came to believe this tall tale.

The brief paper in which Einstein first presented the relativity formula went largely unnoticed. A few physicists, including Max Planck himself, recognized its originality and potential importance. Nevertheless, they weren't immediately convinced that the formula applies universally. That's exactly as it should be. Radical ideas are never instantly accepted in science. It's the job of scientists to be cautious; a healthy skepticism keeps wrongheaded ideas from taking hold. A theoretical physicist's new formula may appear logical, elegant, exciting, thought-provoking, even potentially revolutionary, but only experiments can confirm whether it describes the real world.

The process of verifying or disproving a theoretical idea can be slow and rocky. Already in 1906, an experiment carried out by Walter Kaufmann contradicted Einstein's formulas. But as experience has taught physicists over and over, it's wise to reserve judgment on experimental results until they've been confirmed by independent teams of experts. Working at the forefront of knowledge is difficult, as it usually requires pushing a technological envelope, and errors are not uncommon. Within a year or two, it was clear that Kaufmann's experiment was flawed. It took a full decade before the situation began to settle.

Such steps and missteps are common in science. Scientific knowledge is not born like the Greek goddess Athena, fully grown and armed on day one. Only gradually does the truth emerge and develop its thick, tough armor.

The relativity formula did not appear out of thin air. Various equations that paved the path for Einstein had already been invented by George Francis FitzGerald and by Hendrik Lorentz, and certain key ideas had been introduced by Henri Poincaré. Other authors found formulas similar to Einstein's, but for the wrong reasons. What made Einstein's work so extraordinary wasn't his formulas; it was his conceptual leap. Repurposing the FitzGerald-Lorentz formulas in a way that their inventors had never imagined, he placed them at the heart of a new and radical vision of space and time, foundational aspects of the universe that had been taken for granted since Newton's era. In doing so, Einstein forever changed our understanding of the universe.

Following his new ideas to their logical conclusions, Einstein realized that ordinary objects must carry hidden energy. This brought him to the relativity formula.[4]

7.2 What the Relativity Formula Does Not Say

There are many interpretations of $E = m[c^2]$ to be found in books and on websites, but not all of them are correct. Some describe it as saying that "energy can be turned into matter, and vice versa," or even that "energy and matter are the same thing."[5] But these viewpoints are deeply flawed. For one thing, the letter m stands for *mass*, not *matter*. For another, such a relation couldn't possibly make sense. Energy and matter are in different conceptual categories. The former is a property that objects have, while the latter is a substance that some objects are made from, and they could no more be equivalent than height could be equivalent to bread. Furthermore, we'll see that all matter has energy; it's neither equal to energy nor in opposition to it.[6]

Others claim that the formula says that "energy and mass are the same thing" or that they are "equivalent" or "equal." Such interpretations are not as false as relating energy to matter, but they're not straightforwardly true, either. We've already seen that there are different versions of mass, and as we'll see, there are also multiple versions of energy. If we choose incompatible types of mass and energy, they're not related by Einstein's formula.

And then there's the "=" sign. You might recall that former president Bill Clinton, under questioning by a special prosecutor, said that "it matters what your definition of 'is' is." That remark generated a lot of laughter. But if you're a lawyer, a philosopher, or a physicist, there's merit to this linguistic nit-picking; it's not just a politician's domain.

To get a feel for the subtleties, let's look at a similar (though not exactly parallel) relation. In my right hand, I have a ten-dollar bill; in my left, a stack of a thousand pennies. They're exchangeable, of equal value in an economic transaction. But this does not mean a stack of one thousand pennies is literally the same as, or even equivalent to, a ten-dollar bill. One is metal, the other is paper; they have completely different sizes, weights, masses, colors, flammability, and potential to cause damage if thrown. They are equal only in a particular context and in a particular sense.

Moreover, be careful about what's gone unsaid. Are these American dollars or Australian dollars, and similarly, where do the pennies come from? Ten American dollars are not simply related to a thousand Australian pennies; the exchange rate changes all the time.

But here's a true statement: *if you have D American dollars, you can exchange them for P American pennies, where P is one hundred times larger than D.* Since, like all physicists, I enjoy writing symbols—it makes me look smarter than I am, and it saves space, too—I can rephrase this in an abbreviated form:

$$P = D[100]$$

To say it again: the number of pennies that you will get from a bank, if you hand the teller D dollars, is D times 100.

Despite appearances, this formula does not imply that pennies and dollars are strictly equal in every possible sense. It has a conceptual meaning: American pennies can be exchanged for American dollar bills. And it has a mathematical meaning: it gives a precise instruction for how to convert dollars to pennies using the "exchange rate" or "conversion factor" in brackets.

In a similar sense, the most important aspect of the relativity formula isn't math; writing it in symbols just makes it look that way. Instead, it encodes a relationship between some type of energy and some type of mass. The only role of the *[c²]* part of the formula is as a conversion factor, which we can ignore unless we actually need to convert some E into m or vice versa.

But to interpret the formula correctly, we need to know what type of mass it refers to, and what type of energy. You might think that would be easy. It's rather embarrassing for physics that Einstein's formula, probably the most famous formula in the history of science, is ambiguous. It can be interpreted in two very different ways.

The overarching issue behind the ambiguity is that there are *intrinsic* and *relative* versions of both mass and energy. For each one, there's a separate interpretation of the relativity formula. (You'll be pleased to hear that c is not ambiguous: it always represents the cosmic speed limit.)

In Chapter 5, we encountered both rest mass and relativistic mass, which represent intrinsic and relative versions of the concept of mass. One way to view Einstein's formula is that it relates rest mass to an intrinsic notion of energy; this is the interpretation that we will use throughout this book. A second viewpoint is that m means relativistic mass, while E refers to a corresponding relative notion of energy. (I'll describe these versions of energy in the next chapter.) Although theoreticians in particle physics almost never use the second interpretation,[7] it is often found in other books and in the media, so I'll say a few words about it later.

But even more important than distinguishing these two different interpretations is to avoid merging them! An intrinsic version of mass cannot be equivalent to a relative version of energy or vice versa. Two quantities cannot be simply related if one depends on perspective and the other does not. We risk tripping over our own thinking unless we remain vigilant, always specifying which versions of *m* and *E* we're referring to.

Perhaps you are appalled that scientists, even Einstein himself, have permitted such chaos to take root within their most important equations and concepts. It's true that this disarray often causes trouble for nonexperts, especially when authors don't state explicitly which type of mass they're writing about.

A friend of mine, astonished to learn of these ambiguities in *mass* and *energy*, suggested that physicists might need some adult supervision—perhaps a committee of outsiders to oversee our terminology. Not an unreasonable idea, I agreed. But sadly, no language experts watch over us, and so physics dialect is as messy as the history that has given rise to it. New terms may make sense at the moment that they are invented, but as gaps in knowledge are filled in over time, the original terms often end up seeming inappropriate. Hence names such as *atom* for something divisible, *particle* for something wavelike, *strange quark* for something not strange, and so on. It's easy to complain about the unfortunate choices of the past, but it's hard to undo them.

As a reader, you can use contextual clues to help you guess whether an author is referring to rest mass or relativistic mass. Since speed is relative, any statement that "mass increases with speed" must be alluding to a relative form of mass, such as gravitational mass or relativistic mass. But any claim that the mass of some object has a definite, fixed value—any blanket statement—can be referring only to rest mass. Blanket statements cannot be made about relative forms of mass; such masses vary depending both on objects' motions and on observers' motions.

For example, if an author states that the mass of the electron is smaller than that of the proton, that's a blanket statement that can be true only of rest mass. As individual protons and electrons change speed, their relativistic masses change, too, so the above statement is often false: a sufficiently fast electron can easily have a greater relativistic mass than a stationary proton. (Particle physicists have been pushing electrons to these speeds for many decades.) The same would be true of other relative forms of mass, including gravitational mass.

All electrons are identical, so their rest masses are exactly equal. This allows writers to make blanket statements about "the mass of the electron," referring to all electrons at once. It would make no sense to refer to *the* relativistic mass of *the* electron, since each electron has its own, and what it is depends on how it is moving and on which observer you ask. One can speak meaningfully only about the relativistic mass of a specific electron as viewed by a particular observer.

Yet even though rest mass is intrinsic and remains fixed as an object's speed changes, it can still be modified in other ways. If something perspective-independent happens to the object, its rest mass can certainly be altered. For instance, if a piece breaks off, the rest mass of what remains will be reduced. All observers can see the broken piece and can agree on the rest mass of what is left behind. The change in rest mass, in other words, is itself intrinsic. (In a similar sense, if a loudspeaker's intrinsic volume is increased using the volume control, all observers will agree that the sound has become louder, no matter where they are; the change is intrinsic, not apparent.)

Most importantly for this book, the Higgs field can shift particles' rest masses. When that happens, it's not a matter of perspective. Everyone will agree on the fact of the change, the cause of the change, and the extent of the change.

8

Energy, Mass, and Meaning

In English, the word *energy* is often vague or even mystical. Just wander around online. Auras, a web search reveals, are the "invisible energy fields that surround all living things." You can buy quantum-energy-infused cards and capsules. And did you know that crystals have an extraordinary ability to store, transmit, and transform energy? None of these are what physicists mean by "energy." In ordinary conversation, we speak of people who "sap our energy," describe ourselves as "feeling energized after a good rest," and admire people "with good energy." Even though we can interpret these expressions, their precise meaning is hard to state.

By contrast, energy in physics is precisely defined. Still, it's harder to explain than mass, not because it's vague but because physical objects (or combinations of objects) can organize their energy in many different ways. This makes it impossible to summarize in a few words.

Certain English meanings for *energy* have some parallels in the physics context. If I say, "You don't seem to have much energy today," I mean that you seem inclined to spend the day lying on the couch. If instead you are walking around town with a spring in your step, I might comment on "all the energy you've got this morning." A child forced to sit still for an hour is said "to have lots of energy stored up" and when released and allowed to run around outside is said to be "letting out a lot of energy."

None of these is exactly what a physicist means by "energy," but they aren't far off. Physics energy is both about the capacity for activity—*stored* energy, as you'd find in a car battery or a taut bow—and about activity itself—*motion* energy, as for a moving car or a flying arrow.[1]

Energy is tricky because it can hide. Though both stored energy and motion energy can be visible and obvious, they can also be invisible and easily missed (see Table 1). In daily life, we often fail to recognize where energy comes from and where it goes, leading to misconceptions that are hard to overcome. After Newton understood the basics of motion, mass, and gravity, it took scientists and engineers more than 150 years to figure out how physics energy works.

ENERGY

Motion Energy		Stored Energy	
Visible	Invisible	Visible	Invisible
Plane's flight	Wind	Taut bow	Fuel
Earth's spin	Heat	Raised hammer	Battery
River's flow	Sound's motion	Stretched spring	Gunpowder

Table 1: Both motion energy and stored energy can be in various invisible or visible forms. Energy can be shifted from any one form to any other.

Moreover, the word *energy* is almost a false friend; be careful not to confuse its meaning in physics with its English meaning. For example, in English, it's pretty obvious that a dead chicken has no energy. But in physics dialect, it has plenty. The word *calories* on food packaging refers to the physics energy stored in the food. More generally, physics energy moves from creature to creature as one organism is eaten by the next in the food chain. Each of us stores energy in our bodies so that it's available when we need it, and as we use it and then lose it in our daily activities, we need to replenish it by consuming more food.

But if it's a chain of one thing consuming another, where does the chain start? It begins with plants that obtain energy directly from sunlight via photosynthesis, the conversion of a photon's motion energy into stored energy locked away in biochemical molecules. In turn, that photon of sunlight gets its energy from the Sun's inner furnace, a natural nuclear reactor that can convert protons into neutrons (while producing other subatomic particles). In the end, the energy we obtain from food—and I mean our physics energy, not aura-energy or quantum-energy or crystal-energy—originates with subatomic particles.

8.1 Demystifying Energy

Nineteenth-century physicists and engineers were keen to learn energy's secrets. The industrial revolution was dawning, and the first engines were being designed and built. How much fuel would an engine require? How much work could it do? How could it be made more efficient? These were crucial questions for the technology of the time.

Since engines can both exploit and create heat, it was vital to understand how heat is related to energy. We intuitively sense that they are connected somehow, but it proved challenging to get it straight. Eventually scientists realized that heat involves a hidden form of energy—the motion energy of atoms and molecules, invisibly jiggling and careening in random activity. The appearance and disappearance of heat plays a central role in our understanding of how machines work.[2]

In colloquial English, we might describe a car as follows. The car has energy stored in its fuel. When you press the gas pedal, the engine consumes some of that stored energy to make the car move. Then, once it stops burning fuel—once you take your foot off the gas pedal—the car will begin slowing down. It will soon stop unless you press the gas pedal again and use some more energy.

This makes sense in everyday conversation. But physics energy doesn't work like this. For physicists, the first part is the same: the car has energy stored in its fuel. But rather than consuming the stored energy, the engine converts it into motion energy of the car. The moving car has motion energy simply because it's moving, and that energy came from the fuel; whatever stored energy the fuel has lost, the car has gained it through its motion.[3]

Then, with the accelerator released, the car slows down. But it's not because the energy has been used up. It's because the car's motion energy is quietly being stolen and transformed into an invisible type of energy: heat.

The most important culprit in this secret crime is friction, caused by the rubbing of surfaces against one another as the engine moves, the axles turn, and the wheels press the road. Through this friction, the car's organized motion energy is *converted by dissipation* into disorganized motion energy that our eyes can't see.[4]

As this happens, the total amount of physics energy isn't changing, even though its form is shifting and it is being transferred to new places. Dissipation causes energy to spread out and apparently disappear, but none has been lost; it's all still out there.

A science teacher I know thought up a great analogy: it's like money in bank accounts. You get paid a thousand dollars, so your savings account goes up; that's like filling the fuel tank. Then you transfer the money to your checking account; that's like turning the fuel's stored energy into the car's motion energy. Over the next couple of weeks, the money gradually disappears from your checking account—handed to the bank for your mortgage, stores for groceries, utility companies for water and electricity, and so on. That's like the car's motion energy dissipating and becoming heat in various places. Eventually your checking account is empty, but it's not because the thousand dollars disappeared. It just was something you first obtained and then eventually had to part with; other people have it now. The energy in the fuel isn't gone, but it's not in the car's motion anymore, either; it's gone into heating parts of the car and road, and from there, it's escaped into the wider environment.

This is a brilliant explanation because it reflects the way in which physics energy resembles money. It's precisely measurable—we can figure out exactly how much an isolated object or set of objects has—and amounts flow from one "account" to another, though to follow the flow requires close attention.[5] It's completely different from energy in English, which is often vague and hard to imagine measuring, and which sometimes appears from nowhere or disappears forever.

The fact that physics energy is never lost or gained overall is known as the *conservation of energy*. (This means something very different from what it means in English. Remember that "conservation" in physics dialect really means "preservation.") This gives us a new perspective on the coasting law. Suppose an object is already moving but is isolated from other objects. Since it is isolated, its motion energy can't be stolen because there's nothing to do the stealing—no friction, air resistance, or anything that could cause it to emit light or sound. Its motion energy must therefore remain constant, and correspondingly, it must coast at a constant speed.

8.2 The Secret Within

I still owe you interpretations of the relativity formula in terms of the intrinsic and relative versions of mass. For that purpose, we need corresponding intrinsic and relative versions of energy.

We will see that the energy stored inside an object, which I will call its *internal energy*, is intrinsic to it: all observers will agree about it. But this does not include the additional energy that the object has if it is moving. Since all motion is relative, motion energy must be as well.

Also relative is an object's *total energy*, where we add together its internal energy and its motion energy. Observers will have differing views concerning an object's speed, so they will disagree about its total energy, too. Yet this relative form of energy is important, especially for physicists. The total energy of an isolated object (or of an isolated set of objects) is conserved, while this is generally false for both motion energy and internal energy separately.[6]

Let's now employ our cell phones to explore the notions of internal energy and total energy more carefully. When I charge my phone each night, energy originating in a power plant is carried by electrons through the charger's wire. Inside the phone, that energy is converted and stored in the battery's chemicals, increasing the phone's internal energy. This represents an intrinsic change to an intrinsic property; all observers agree that my phone is now fully charged.

Then, as I use the phone during the day, making calls, taking photos, and watching videos, the stored energy gets converted into the motion energy associated with light, microwaves, and heat, and into electrical energy that the camera needs to capture and store images. The internal energy in the phone decreases, and soon the phone needs to be recharged.

One day I take the phone, 50 percent charged but turned off, on a flight across the continent. From your perspective, watching my plane fly overhead, the phone is moving, and so it has motion energy as well as internal energy. Where did the phone's motion energy come from? From the plane's jet fuel, which was used to power the movement of the plane and of everything inside it. Notice, however, that the burning of jet fuel put no energy inside the phone; you can't charge a device simply by making it travel quickly. So even though you see the phone as now moving at a fast clip, you know that its internal energy remains that of a half-charged phone.

From my perspective, however, the phone is stationary, so it has no motion energy; its total energy is the same as its internal energy. Yet I agree with you that its internal energy is that of a phone with a half-full battery. Since you see the phone as having motion energy and I do not, you view its total energy as larger than I do. In other words, both its motion energy and its total energy are relative.

By contrast, the phone's internal energy is intrinsic to it. You and I concur that it is equal to the energy of a phone whose battery reading is 50 percent. We both agree that if I turn the phone on midflight, I'll be able to watch two TV episodes on it, but not three, before the battery runs out. (Whether I was able to watch the third episode of a series can hardly be a matter of perspective! Imagine the conversation that might ensue if it were.)

Thus, it's only internal energy, being perspective-independent, that can play a role in rest mass. Motion energy, being relative, cannot. This is summarized in Table 2.

	Intrinsic – Independent of speed – Observers agree	**Relative** – Depends on speed – Observers disagree
Meaning of *E*	**Internal Energy**	**Total Energy** – Internal energy plus motion energy
Meaning of *m*	**Rest Mass** – Intransigence when seen as initially stationary	**Relativistic Mass** – Intransigence when pushed along direction of motion

Table 2: The two meanings of *E* and *m* relevant to Einstein's relativity formula.

But our focus on the energy stored in the phone's battery was just a teaser. The phone has far more internal energy than that. This is what Einstein revealed through the relativity formula. To interpret it, I'll rewrite it in a simple way by (1) moving the *[c²]* (using division) from the *m* side to the *E* side and (2) switching the two sides. That gives us

$$m = \frac{E}{[c^2]}$$

In this form, Einstein's formula makes its point most clearly.

Let's interpret *m* as an object's rest mass—its intransigence when stationary. And let's correspondingly interpret *E* as internal energy—the amount of energy stored within the object. Then the relativity formula says the following.

First, it has a conceptual meaning: an object's rest mass is a measure of how much internal energy it has. In other words, *m* (rest mass) is secretly *E* (internal energy).

Second, it has a practical, precise meaning, which you will need if you actually want to convert *E* to *m* or vice versa: an object's rest mass is equal to its internal energy divided by a conversion factor, namely, the square of the cosmic speed limit.

Now we see where Newton, chemistry class, and the dictionary definition of mass all go wrong. Instead of the *quantity of matter* inside, an object's rest mass is the *quantity of energy* inside. It's not that energy and mass are the same thing, or equivalent in general. It's more specific than that: the rest mass of an object arises from the energy stored within it.

There's a simple reason why this wasn't known before Einstein's time. The majority of the internal energy in ordinary objects is stored in an invisible, hidden form, which we will explore throughout the rest of this book. It was only in the late nineteenth century that scientists began to stumble on its hiding places.

Since every electron has the same rest mass, we have now learned that every electron has the same amount of energy stored inside it. No energy ever flows in or out to increase or decrease that internal energy, either. It's completely unlike a car or cell phone. Why is this the case? Where does an electron's internal energy come from, what does it consist of, and why can't it vary? We'll spend a lot of time figuring that out.

The lesson of the relativity formula is that intransigence, the stubbornness that resists changes in motion, comes from energy. It's the energy stored inside an object that makes it harder to throw or catch.[7]

But now perhaps you are wondering about your cell phone. When you charge it, does its rest mass really increase? Yes, it does, a tiny bit. And when you use your phone, its rest mass goes down a tad. It's true of computers, too. But the amount of energy flowing in and out of a charger is so small, compared to the amount of energy already stored in the atoms of the phone or computer, that you'll never feel the difference. Nor will you detect the increased rest mass of a rock that's been sitting in the sun, absorbing energy from sunlight.

Most of the internal energy—most of the rest mass—of any ordinary object is stored in its protons and neutrons. What goes in and out of a cell phone when you charge or use it alters its rest mass by less than one part in a billion. You, too, are losing some of your internal energy as you give off body heat, but you'll never be able to shed much mass that way.

What a curious world it would be if these changes were big enough that we could feel them! Imagine if a hot ball were more difficult to throw than a cold one, or if lifting a cell phone were more difficult when it was fully charged, or if we could lose mass simply by moving to a colder climate. Our common sense about how the world works would certainly be different!

Before moving on, let me say something about the second interpretation of the relativity formula that I alluded to. We get it by taking E to be *total* energy and m to be *relativistic* mass. In this view, an object's relativistic mass, its intransigence when it's already moving and you try to increase its speed, is its total energy divided by c^2. Both total energy and relativistic mass increase with speed and depend on an observer's perspective, so this gives us a different but consistent interpretation.[8]

Ironically, most particle physicists don't use either of these two simple interpretations of the relativity formula. We use a third one![9] Who knew there could be so many ways of parsing three letters, one number, and an equal sign?

9

That Most Important of Prisons

Knowing now that rest mass is secretly internal energy, it's natural to ask, "What is the internal energy that provides a human's rest mass, and where is it stored?" Before we get into the details, let's first reflect on how much energy is in there.

The first person to compute the energy of a human was Einstein himself. Having guessed correctly that his relativity formula was general, applying not only to certain elementary particles but to all objects in the universe, he could calculate what was stored in his own body, even without knowing how it got there or where it is located. I do not know of any story concerning his thoughts after he first did the calculation. But at the end of his article on the relativity formula, he informed the rest of the world of the answer, hidden implicitly in a single terse sentence.[1]

To put the answer in context, let's first envisage the energy needed for a human just to breathe and walk around normally. Amusingly, the energy demands of a human body are about 60 watts—the same as a single old-style incandescent bulb or a few modern LED lightbulbs.

But this energy for daily consumption, obtained from food, is minuscule compared to the energy stored in your rest mass, mainly within your protons and neutrons. So let's imagine that your body, instead of relying on food, obtained its energy supply by transmuting the energy stored within some of your protons and neutrons. To provide the 60 watts of power that you would need, the relativity formula tells us, you'd have to convert the rest masses of a trillion (1,000,000,000,000) protons and neutrons every second.

That might seem like a lot. But it's tiny compared to your body's stash of protons and neutrons, which number 10^{29}, *a hundred thousand trillion* trillions.[2] So if it takes a trillion protons to power a 60-watt bulb for one second, the energy stored in your body could keep that lightbulb lit for about *a hundred thousand trillion* seconds (10^{17}). That's about ten billion years, roughly the age of the known universe.

If we could extract your energy more quickly, then instead of using it to light one bulb for ten billion years, we could light ten billion bulbs for one year. Ten billion is roughly the number of bulbs in the United States, or slightly more than the number of people alive today. For a full twelve months, the energy in your body could illuminate a superpower, or supply all the caloric needs of the world's entire population.

If that seems impressive, consider the consequences of releasing all that energy suddenly, setting it free and converting it into the motion energy of photons, electrons, and other particles. The resulting explosion would be equivalent to a gigaton—one billion tons—of TNT explosive. That's 100,000 times more violent than the nuclear explosions that destroyed the cities of Hiroshima and Nagasaki. It's a thousand times more powerful than a typical hydrogen bomb in the arsenals of the United States, Russia, and China; in fact, it exceeds tenfold the largest nuclear weapon ever tested.[3] It's larger than the volcanic eruption that destroyed the island of Krakatoa in 1883, itself more potent than the similar blast in 2022 within the Republic of Tonga. Only the 1815 eruption of Tambora, possibly the largest volcanic explosion on Earth in the last two thousand years, affecting the Earth's climate for months and causing widespread crop failure and famine, may have been greater than the explosive potential of a single adult human being.

Just consider, for a moment, how much energy we carry in our species, eight billion strong. It's far more than the motion energy of the meteor that killed off the dinosaurs. In fact, though it's not enough to blow up the Earth, it's enough to melt its solid crust entirely.

As you contemplate the staggering amount of energy stored in a human body, remember that we can't do without it. If I lacked the energy to destroy a megacity and all its suburbs, I'd blow away like packing foam or a dead leaf, unable to survive and function in the world. Our deadly capacity is the unavoidable price for functional bodies and brains—for intelligent life on Earth.

You might well wonder where all that energy came from. How did each of us end up with a gigaton of TNT stored in our bodies? This is a question I will address in the book's last chapters.

Most fortunately, despite the fact that we're nuclear gunpowder, we're not at risk of detonating. To make a nuclear explosion requires a chain reaction, in which each disintegrating atomic nucleus causes others around it to disintegrate. You need special radioactive materials to start and maintain a chain reaction: uranium, plutonium, or something similar. For many reasons, it's a good thing we're not made from such stuff. In short, we're bombs only in principle. There's no way to set us off.

This is somewhat comforting. And yet we would probably be better off without this knowledge. A species that combines curiosity and ingenuity with deviousness, brutality, and a tendency toward irrational hatred is already incredibly dangerous to itself and to everything around it, even without the ability to turn rest mass into explosions.

That leads me to a brutal, little-known double entendre. You've surely heard nuclear bombs referred to as *weapons of mass destruction*. But you may not have realized that this is literally the case. It is the destruction of rest mass that drives the explosion, in which conservation of energy allows internal energy of atomic nuclei to be converted to the motion energy of flying subatomic particles. As internal energy is transformed into the energy of human violence, a small amount of rest mass is lost from the universe.

While we tread gingerly under the threat of nuclear annihilation, the universe, vast and unconcerned, continues as it has for billions of years, blithely exploiting rest mass to destroy and create stars, planets, and their denizens, if any. The energy hidden in atomic nuclei fuels the lights that illuminate the void and powers supernovas that spread the innards of dead stars across their galaxies. Will the fate of the human species even merit a footnote?[4]

Setting such existential questions aside, let's turn our attention back to the rest masses of protons and neutrons, which host the vast majority of the energy in material objects. What is the nature of their internal energy, and how does it come about? Spoiler: the Higgs field has rather little to do with it.

Our tour of the subatomic realm revealed that the proton is full of particles—quarks, anti-quarks, and gluons, all pulling on each other via the strong nuclear force, as shown in Fig. 16 (p. 92). The pulls are so strong that they whip the particles around at or near the cosmic speed limit, drag them back in if they try to flee, and slam them repeatedly into one another.

Whereas atoms are elegant ballrooms, protons are chaotic, tumultuous dance floors. No dancer ever escapes, either. The strong nuclear force is so overwhelming that you will never find a quark or gluon on its own; each is permanently trapped inside a proton or a neutron.[5]

This trapping effect, often called *confinement*, lies behind several striking features of the proton and neutron. In particular, it explains how the proton's rest mass can be so much larger than the rest mass of its parts.

According to the relativity formula, if an object contains internal energy—energy that remains always inside and travels along with the object—then that energy contributes to the object's rest mass. It doesn't matter what form that internal energy takes.

That's crucial for the proton and neutron. The rest masses of up and down quarks (and anti-quarks) are small, less than 1 percent of a proton's. Gluons, like photons, have zero rest mass. The reason that a proton's rest mass is much larger than the sum of its quarks' rest masses is that the particles whizzing around within it carry a lot of motion energy. This motion energy never escapes, since these particles are trapped, and so it is permanently internal to the proton. Even when a proton is stationary, with no overall motion energy of its own, the substantial motion energy of the particles inside it is ever-present, counts as *internal* energy, and contributes to its rest mass.

The energy required to hold the proton together provides an additional source of rest mass. This stored energy, also arising from the strong nuclear force, is similar to the energy stored in a stretched spring or in the rubber skin of an inflated balloon. When a balloon's nozzle is released, the energy stored in the stretched rubber is converted to motion energy of the air inside, causing it to deflate rapidly. Similarly, there is energy involved in keeping the proton's speedy particles from escaping their enclosure; if you could turn off the strong nuclear force, or somehow poke a hole in a proton's outer skin, the proton's contents would go flying out. Because that stored energy travels with the proton wherever it goes, it, too, counts as internal to the proton and contributes to its rest mass.

By keeping the proton's contents trapped and in rapid motion, the strong nuclear force both maintains the proton and bears responsibility for most of its rest mass. Only a small portion of the proton's rest mass arises directly from the rest masses of the particles within. This gives us an explicit example of rest mass being gained. We start off with quarks and anti-quarks that have little rest mass and gluons that have none, and yet we end up with a proton or neutron that has considerably more. Compare this to atoms, where we have almost the opposite situation: an atom gets almost all its rest mass directly from the rest masses of the particles inside it, namely, its electrons, protons, and neutrons.

When it comes to rest mass, the whole can be greater than the sum of its parts. (It can also be less than the sum of its parts, as is the case in many familiar contexts.[6]) Since most of our rest mass is in our protons and neutrons, we, too, are greater than the sum of our parts, thanks to the relativity formula.

At the time of the Higgs boson's discovery, it was often stated by journalists that the Higgs field gives everything in the universe its mass. This, too, is a phib, a bad one. The Higgs field gives quarks and anti-quarks their rest masses, but it doesn't provide us with much of our own. Any ordinary object obtains the majority of its rest mass through the efforts of the strong nuclear force.[7]

In fact, even if there were no Higgs field, so that quarks' and anti-quarks' rest masses were zero, protons' and neutrons' rest masses would still be substantial, hardly shifted from what they are today. In other words, protons with rest mass can be built entirely from particles that have none! This is possible whenever a trapping force is combined with the relativity formula.

Once particles are imprisoned, their motion energy is trapped, too, and contributes to their prison's rest mass even if they have no internal energy of their own.[8]

"An interesting tale," one of my adult non-science students remarked, "but how do you know all this? You surely can't see inside a proton to check, can you?"

Definitely not! Both experiments and computer studies were needed. The formulas for the strong nuclear force were uncovered in the early 1970s, and in the next few years, physicists who studied the formulas developed the modern conception of the proton and neutron. The details were debated, but eventually computers became fast enough to carry out a realistic simulation of a proton. These simulations, along with measurements of protons' innards at particle accelerators, have confirmed that our basic understanding of protons (and neutrons) is well-founded.

Seeing how unimportant quarks' rest masses are, you might reasonably conclude that the electron's rest mass is equally unimportant. After all, electrons contribute only a fraction of a percent to an atom's mass. But you'd be wrong. The reason has everything to do with trapping, or the lack of it.

As it traps quarks inside protons and neutrons, the strong nuclear force sets the size of their prison. The extent of the proton's core would barely change even if the quarks' rest masses were reduced to zero. By contrast, the electric forces holding electrons inside atoms are much weaker and cannot trap them. This has an important consequence: with smaller rest masses, electrons could more easily stray from their atomic nuclei, making atoms larger (Fig. 17) and much more fragile. If we somehow made the electron's rest mass a thousandth of what it is today, atoms would grow a thousandfold, becoming so flimsy that you and I would evaporate away even at room temperature.[9]

Shrinking the electron's rest mass slowly down to zero, we'd find that even in the cold of outer space, all electrons would escape from their nuclei. As particles of zero rest mass, they'd go sailing off into the universe at the cosmic speed limit, much like starlight. Atoms would completely disintegrate, and everything made of ordinary material would vanish in a puff of subatomic "smoke."

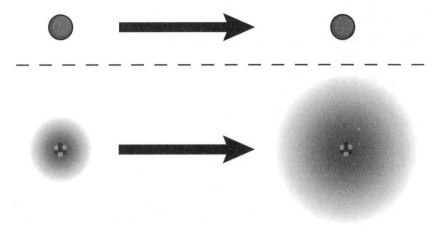

Figure 17: (Top) If quarks' rest masses were smaller, protons would hardly change. (Bottom) If the electron's rest mass were smaller, atoms would grow and become more fragile.

If instead the electron's rest mass dropped to zero in an instant, the impact would be more spectacular: you and I and all other ordinary objects, including Earth, would explode. The detonation would pale compared to a thermonuclear blast, but it would still heat our planet and its creatures far above a survivable temperature.[10]

We are truly dependent upon the electron's rest mass, small as it is; were it to disappear, nothing in or on Earth would last an eyeblink. This, in turn, highlights our secret reliance upon the Higgs field. If it didn't exist, or if it hadn't switched on so it could play its important role, atoms would never have formed.[11]

This is not the only problem that an absent or malfunctioning Higgs field would cause. Though changes to protons and neutrons might be minor, impacts on certain atomic nuclei, some of which are surprisingly delicate, could be catastrophic. I won't dwell on the details, though. It hardly matters which nuclei survive if electrons can't combine with them anyway.

The electron, we might say, is the tail that wags the atom. This focuses our attention on the origin of its small but crucial rest mass.

As we began to address this question in class, one of my students, a retired doctor, sat back in her chair with a skeptical look. "So, since electrons get their rest mass from the Higgs field, doesn't that mean the Higgs field has to somehow give each electron some internal energy?"

When I affirmed, a bit hesitantly, that it does, her look shifted to one of quiet triumph. She'd caught me in an apparent contradiction. "But you also told us that as far as anyone knows, an electron has no size. This sounds like nonsense: How can an object with no size possibly have energy inside it?"

It does seem inconsistent. How can there be energy within a particle that has no interior? Even if you resolved that puzzle, more mysteries lie ahead. For instance, how can the Higgs field assure that every single electron in the universe gets exactly the same amount of internal energy, an amount that remains constant over billions of years?

We cannot hope to find answers, and the principles that underlie them, without a clear conception of what an electron is, what a field is, and why the two can have something to do with one another. As physicists learned in the middle of the last century, electrons are quite different from the dots often drawn to represent them, and their internal energy is of a sort and stored in a fashion that you could never guess from the atomic cartoon's little ball. Instead, we must understand the cosmos in a surprising way: as a musical instrument, with an electron as a quantum tone.

WAVES

Every physics professor has colleagues who tape nerdy cartoons to their office door. "Particles, particles," reads the caption of one, in which a janitor in a physics laboratory is wearily sweeping up dust that lies scattered on the floor. But if electrons and quarks were actually like grains of dust, I don't think I'd have become a particle physicist. Reality is so much more interesting than that. In fact, the particles that make up our bodies and the objects around us are also waves.

This last statement sounds more like mysticism than physics. Knowing what we do about waves from ordinary life, it's hard to imagine what it could mean. Our most direct and tangible encounters with waves are in water—at beaches, in a bathtub, or on lakes and rivers. We also see waves in a cup of liquid when we jostle it or blow on it. Other familiar waves include those that move down a piece of rope or a long piece of cloth when we shake one end. It's from these examples that we build our common sense about waves. They move. They slosh. They spread out. They're squishy rather than hard. They occur within stuff that you can see or at least feel, such as water or air or fabric or rock. And they don't last forever; if you make some waves yourself, they're soon gone.

In secret, waves play a much larger role in daily experience. We aren't intuitively aware that light and sound are both waves; our brains don't clue us in. Their wavy character is something we learn from a book, a science museum, or a classroom. That's why light's curious ability to travel all the way across the universe doesn't immediately trouble our common sense, either as adults or as children.

The idea that we ourselves are made from waves is confusing at best. We never encounter ordinary objects that are obviously made out of waves. How could you actually construct a stable structure using waves in water or air? Waves aren't like bricks or wooden boards that can be counted and organized and arranged and attached to one another, with the potential to stay fixed in place for years or even centuries. When we're told that electrons, protons, and neutrons are particles, that makes intuitive sense; we imagine them as though they're little balls that can be stacked and assembled into atoms, molecules, and ever more complex structures. But waves? How can nature make a table, a tree, a hand, even an entire planet out of things that move and slosh, are squishy, and are soon gone?

Again, our intuition impedes our comprehension. We first need more insight into waves and a sense of how the universe is similar to a musical instrument. Then we will have to confront other questions. What are these waves made of? What is the nature of the empty space that the waves move through? And what's unique about microscopic waves? The counterintuitive answers will lead us toward an understanding of how waves of an unfamiliar sort can be the building blocks of a human being.

= 10 =

Resonance

Some years back, on a summer afternoon, I was chatting with a friend as we sat on a bench at a little playground. His wife and son were at the swing set. The boy wasn't yet old enough to propel himself, so his mother was giving the swing a regular push.

Her phone beeped, and she turned aside in order to send an extended reply to a text message. As she typed, friction took its toll, and though the boy continued to swing, he gradually lost altitude. As he came almost to a stop, he wailed, and his mother put her phone away and went back to her parental obligations.

As we watched the child in his hypnotic motion, forward and back, forward and back, I had a sudden thought, which I expressed in oracular terms. "Right there," I said. "That's the secret to the universe."

"Whatever are you talking about?" asked my friend, looking over at me with an amused smile.

I asked him how his wife knew when to push.

He raised his eyebrows quizzically. Perhaps he was wondering whether I had any common sense.

"See, it's all about *resonance*," I said. "And so is the universe."

Whether you've been in the role of child, parent, or both, you know: a child's swing must be impelled rhythmically, at just the right times. Precisely when the swing comes to a stop on its way back, that's when you give a shove. In other words, *you let the swing itself tell you how often you should push*. It knows. If you do anything else, the swing won't fly high and its motion will be irregular and jerky, to the displeasure of the child.

This, for a change, is a bit of common sense that survives scientific scrutiny. What's behind that common sense is an intuitive understanding, at least in this context, of resonance. (Similar intuition governs other activities, such as rocking a stuck car back and forth when you're trying to extract it from the mud.) Resonance is central to the workings of swings, of clocks, of musical instruments, and of our universe as a whole.

Though it may not be obvious why, in a book on elementary particles, a discussion of resonance immediately follows a discussion of mass, it's no accident. The precise connection between resonance and rest mass is one we'll be edging toward for many chapters to come.

To understand resonance better, we need to explore the basics of vibration—of back-and-forth motion as well as other forms of back-and-forth change. Many of our most common experiences with vibration are musical. If you've played a guitar or any of its cousins—lutes, sitars, ouds, kotos, ukuleles, violins, cellos, double basses, zithers, dulcimers—you will already have some valuable intuition. Even if you haven't watched closely as one was played and you don't have immediate access to such an instrument, you can teach yourself all the necessary lessons using a single piece of string. (A thin string perhaps a yard or meter long will do fine. Tie one end around a fixed object, such as a doorknob or post, and wrap the other end tightly around something heavy but movable, such as the top of a wooden chair. Then move the chair until the string becomes taut enough that it makes an audible musical tone when you pluck it. If you don't hear a tone but can easily see the string vibrating back and forth, it's not taut enough; move the chair to increase the string's tension and try again.)

When you pluck a string, it begins vibrating. You may not be able to discern this visually; typically, if the vibration is audible to the ear, the motion is too rapid for the human eye to follow. Nevertheless, the back-and-forth motion will make the string appear blurry. Over time, the blur will narrow and the sound will become softer. Eventually the string's appearance will become crisp again, and the sound, correspondingly, will cease.

When a string is plucked, a flute played, or a bell struck, we hear a sustained sound, characterized mainly by two features. It has a *pitch*, or *note*, or *tone*—three words that I will use almost interchangeably here.[1] And it's heard with a certain particular *loudness*, or *volume*.

That's what our brains perceive, anyway. But the world is full of illusions that fool our brains, so we should be cautious. What's actually going on?

As a physical process, a vibration is also characterized mainly by two features. First, there is a certain rate of back-and-forth motion. The number of full cycles (from back to forward and back again) that occur each second is called the vibration's *frequency*. A typical child's swing has a frequency of about one-third of a cycle per second, meaning that it makes a full cycle in about three seconds. Meanwhile, the first string on a guitar, the one farthest to the left when we face the instrument, has a frequency of about eighty cycles per second, too fast to follow by eye but in the right range for us to hear.

Second, there is the degree or size of the back-and-forth motion, which is called the *amplitude* of the vibration. The harder the parent pushes the child, the larger the amplitude of the swinging. If the parent, distracted by a cell phone, stops pushing, the amplitude begins gradually to decrease. Similarly, a string plucked firmly will have a larger amplitude than one plucked gently; over time, its amplitude will steadily diminish.

Where vibration is concerned, frequency is *how often*, while amplitude is *how far*. And unless you're interested in more subtle details, such as *timbre* (which makes a violin sound different from a piano even when they play the same note), there's not much else to know.

Remarkably, these two properties of the vibration largely determine what we hear: frequency sets the note, and amplitude sets the volume. (More precisely, amplitude determines the sound's intrinsic volume, while the apparent volume also depends on how far we are from the guitar.) As the amplitude of the vibration dies away, the volume of the sound does, too.

This is a lucky break for science: our experience of sound corresponds closely to what is actually happening, which isn't true for our other senses.

It's no accident that ancient Greek scholars understood sound better than most other everyday phenomena, recognizing that human hearing is related to vibration.

More (i.e., larger) amplitude means more (i.e., louder) sound. More (i.e., higher) frequency corresponds to a higher note—a higher-pitched sound. The notions of high and low notes, or high and low pitches, are standard musical terminology that have entered idiomatic English, as in the sentence "She hoped to end her career on a high note." (If you already know this terminology, you can skip the rest of this paragraph.) We understand it intuitively from singing. "High" notes feel high in our throats; they are the squeakiest sounds we can make. "Low" notes feel lower in our bodies, as when we imitate the roar of a lion. Children and most adult women are able to sing higher notes than most adult men, and boys' voices change from high to low pitch when they pass through puberty.

On any instrument with multiple strings, the strings make different notes because they vibrate with different frequencies. The sixth string on a standard guitar vibrates four times as fast as the first one, and with that higher frequency, the string makes a note that we describe as "higher." On a piano, the keys to the far left make "low notes," while those to the far right make "high notes," with correspondingly low and high frequencies.

Now, here's an important question for this book. After you pluck a string, the sound gradually dies away—the loudness decreases—but the pitch meanwhile stays constant. Since amplitude corresponds to loudness and frequency to pitch, this means that *as the amplitude of a plucked string's vibration decreases, its frequency does not change.* Why?

Whatever the answer, it's vital for music and musical instruments. If it weren't true, a plucked string wouldn't produce a pure note; the pitch of the sound would be unsteady, like a siren or a yowling cat. Perhaps you've heard the singing of Bob Dylan from the 1960s? A great musician. But because his style involved sliding the pitch all over the place, it was almost impossible to sing along with him. Now imagine if your guitar were always like that. (Dylan's wasn't.)

Not only that, if frequency changed along with amplitude, playing instruments would be much more challenging. Most instruments allow the player to take one action to choose which note to play and a separate action to determine the loudness of the sound. On a cello, the placement of the left hand's fingers fixes the note, while the pressure of the right arm on the bow determines the volume of sound. On a piano, the choice of which key to strike selects the note, while the volume is set by the striking force. Such a separation into two distinct actions would be impossible if frequency and amplitude were interdependent. On a guitar, a player aiming at a particular note would have to pluck the string with exactly the right amount of force.

Fortunately for instrument makers and musicians everywhere, these potential horrors are just imaginary. That's because of a wondrous feature of resonance. Many vibrating objects have a *resonant frequency* (or *resonance frequency*). When initially disturbed and then left alone, such an object will always vibrate at that special frequency, no matter what the vibration's amplitude.[2]

In this respect, frequency is rigid, while amplitude is not. Amplitude can easily be changed by plucking, striking, or blowing an instrument with greater or lesser effort. It's a natural, easy thing; a toddler can do it. (As any parent will confirm, young children can always figure out how to make something sound louder.) By comparison, changing the frequency of a string or other instrument is usually more difficult, requiring cognition and physical dexterity beyond a toddler's capacities.

Because of resonance, a string's frequency is independent of how it has been perturbed. The same notes will sound whether guitar players pluck their strings with their fingers or with a pick. Violinists can play identical notes by drawing a bow across their strings or by plucking them. A piano's eighty-eight sets of strings produce the same eighty-eight notes whether they are struck with the hammers connected to the piano's keys, scraped with a credit card, or plucked. Although the character and loudness of the sound will depend on how the string is played, the note produced never changes.

In a sense, frequency and amplitude keep out of each other's way. This carries over to our perceptions: frequency determines pitch without affecting loudness, while amplitude determines loudness without affecting pitch.[3]

This is why resonance is the foundation for most instruments that can make a steady, sustained tone: bells, guitars, violins, pianos, pipe organs, flutes, trumpets, xylophones, and many more. It's resonance that makes playing the instrument predictable and its notes reliable and steady. Without resonance, making music would be dramatically more difficult.

Making humans would be dramatically more difficult, too. In a very real sense, *resonance underlies the entire cosmos*, dictating properties of the elementary particles out of which we and everything else are formed. It is here that we find parallels between our universe and a musical instrument.[4]

Still, I don't want to suggest that the universe makes music in the ways that we do. A guitar, with a diverse but organized set of pitches available, is ideal for making complex melodies and harmonies that can convey human mood and emotion. The universe's "music" is more limited; in certain ways, the universe is far less flexible than a guitar. But as we'll see, its peculiarities make it a much better place to live.

"Does your guitar analogy have something to do with string theory?"

My colleague, a social scientist by vocation and a musician by avocation, had recently read a book about the subject of string theory, which (in its most ambitious form) seeks to provide a complete understanding of the universe's inner workings—to explain all the basic particles and forces, leaving no missing pieces.[5]

"No, string theory is something else altogether," I replied. "The analogy here is different. Roughly speaking, whereas the guitar has vibrating *strings*, the universe has vibrating *fields*."

The universe's fields, from the electric and magnetic fields to the Higgs field, have been extensively studied in experiments, and we know a great deal about them. String theory, sitting at the next level of potential knowledge, represents an attempt to explain where the fields come from.[6] So far, no one has yet discovered a way to test the idea in an experiment, and so for now it remains speculative. Since our focus here is on what is known or is knowable soon, I won't say much about string theory, or other speculative theories, in this book.

"So would you call that *field theory*?"

"Exactly right," I acknowledged. "The math used to describe fields and particles is called *quantum field theory*."

Somewhat as a guitar has strings that reach across it and that vibrate when disturbed, the universe has fields that stretch across it—everywhere, in all directions—that similarly can vibrate when they are disturbed. I've written "somewhat" because the analogy has a flaw, which I'll repair a few chapters from now. But the essence of the analogy holds true: the universe resembles a musical instrument.

10.1 The Flow of Energy

Without the push of a parent, a toddler on a swing loses amplitude. The sound of every guitar string slowly dies away. When we bump a bucket of water, little waves appear on its surface, but not for long. In daily life, vibrations come and go, never lasting for weeks, years, or universes.

Though this might seem obvious and normal, it's another bit of faulty common sense. It is merely the vibrational analogue to the resting law, the one that incorrectly claims that moving objects eventually stop.

Like steady motion, a vibration does not degrade on its own. It degrades because its energy is being stolen. Just as dissipation steals the motion energy of a car, slowing it when the gas pedal is no longer pressed, it also steals the energy of a young child on a swing. As the energy available shrinks, so does the vibration's amplitude.

If there were no dissipation, an object could vibrate forever, just as it could coast forever. We never see this happen in ordinary objects. Yet vibration without dissipation is common in the extreme microscopic world. In fact, we and the objects around us are formed from vibrations that can continue indefinitely, as there's no friction or other processes that can dissipate their energy. We will see how this works in a later chapter.

Nevertheless, there is no vibrational analogue to the principle of relativity. While steady motion feels the same as no motion, you can't mistake regular vibration for its absence; they feel very different.

For a child's swing, dissipation is mostly due to friction, the rubbing of the chain against its support. It steals the energy of the swinging child and turns it into heat. To keep the motion from dying out, the parent needs to keep pushing, adding energy to replace what friction is taking away.

But dissipation for a guitar string is much more interesting (Fig. 18). When you pluck the string, you're putting energy into it; without that energy, it couldn't vibrate. As its vibrational energy gradually dissipates, only a fraction goes via friction into heat. Much of it goes into vibrations of the guitar as a whole and from there into sound—into vibrations of the air, which take the form of moving ripples. These *sound waves* travel outward, carrying motion energy that they took from the vibrating guitar. When listeners' eardrums vibrate from the sound, the energy that powers that vibration originated in the guitar player's finger, transferred through the guitar to the sound waves and out to the audience.[7]

Figure 18: The player plucks a string, adding energy so that it
vibrates. Sound waves carry off some of this energy and transfer it
to listeners' eardrums, making them vibrate, too.

This flow of energy is central in music. It's all well and good to create a vibration on or in an instrument, but if dissipation didn't convert the vibration into sound waves, and sound waves couldn't make eardrums wobble, no one would ever hear it. Such transformations from one type of vibration to another have many analogues, both in general and in particle physics. This is something we will return to.

As the energy from the player's plucking finger gradually dissipates into heat and sound, there's less available for the guitar string, so the amplitude of its vibration shrinks. This in turn assures that the sound waves produced by the string lose amplitude, too, and so listeners hear the sound becoming softer. Eventually, after all the energy's been carried away, quiet is restored.

10.2 The Pendulum and the Universe

A child's swing; a ball at the end of a cord; a key on a long chain. Each is an example of a pendulum, a simple object that displays resonant vibration and dissipation. It also provides essential insights into the workings of the universe.

Historically, an important application of the pendulum was as the basis of a clock. Prior to the seventeenth century, the best clocks were powered by a wound spring; as the spring uncoiled, it drove a gear mechanism that counted off the seconds. However, this method was less than ideal. At first the clock would keep time, but as the spring unwound, it couldn't push the gears as well, so the clock would run slow.

The pendulum clock solved this problem. A pendulum makes a good clock for the same reason that a guitar string makes a clear and steady musical tone. Unlike a spring-driven clock, which counts time more slowly as it winds down, a pendulum clock, as it loses energy, swings less widely but no less often. It's resonance that underlies this reliability. Galileo himself, as a young student, recognized this property of a pendulum, and a few decades later, Huygens designed a practical clock based upon it. From then until the 1930s, the pendulum clock remained the most precise timekeeper available.

To confirm this for yourself, take any sort of pendulum, pull the hanging object (the *bob*) a little to one side, and let go. The bob will oscillate at a certain frequency (Fig. 19). Now stop the bob and restart it with a larger amplitude. You'll see that, despite more dramatic motion, its frequency is the same.

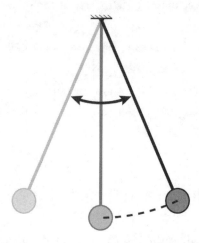

Figure 19: The frequency of a pendulum is how often it swings; its amplitude is how widely it swings (dashed line). The two are independent (as long as the amplitude is not too large).

Over time, the amplitude will decrease due to dissipation from friction, but the frequency won't change. You can counter the dissipation, if you want, by adding a little energy now and then, most efficiently if you push the bob regularly with the same frequency as the pendulum. That's what parents do to keep their swinging children in motion.

The pendulum's frequency remains the same whether you pull the bob, tap it with a hammer, or bump the support it hangs from. Nothing makes any difference. The amplitude is up to you, but the resonant frequency is a property of the pendulum itself. It's intrinsic to the pendulum and out of your control.

I learned this in third grade. My teacher attached an apple to a string and made it swing in many different ways. At one point, he even cut the apple in half. We were all lulled into complacency; no matter what he did, the frequency never changed. Then, at the last minute, he surprised us.

As it turns out, it's often not that hard to shift a vibration's resonant frequency, as long as you know what you are doing. I'll tell you what my teacher did in a moment. But let's focus on guitar strings, whose pitch can be changed in several ways.

1. A guitar is usually played by plucking or strumming strings with the right hand and placing the fingers of the left hand on the neck of the instrument. It's the left hand that's responsible for the guitar's flexibility; without it, the six-stringed instrument could make only six notes. The fingers of the left hand are used to *shorten* the strings. Placed on a string and pushed against the neck, a finger divides the string into two segments. When the right hand plucks that string, only the lower segment vibrates, as in Fig. 20. The shortened string vibrates with a higher frequency, and thus produces a higher note, than the string at full length. The same idea applies to many other string instruments, including the guitar's many cousins and the violin family. Different degrees of shortening give different notes, providing the instrument with the versatility needed for complex music.

Figure 20: (Left) A guitar, showing one string. (Center) If plucked, the
full-length string will vibrate at the string's resonant frequency. (Right)
If the guitar player shortens the string by placing a finger on it, the resonant
frequency increases, and the string produces a higher note.

A similar principle applies to wind instruments such as flutes and saxophones. Inside a flute is a column of vibrating air. If the player's fingers are pressed on all the flute's holes, the air column is as long as the instrument. As the player lifts fingers, the holes let air escape, and the column becomes shorter, causing it to vibrate with a higher frequency. That's how a wind instrument can produce many different notes despite having a fixed overall length.

Similarly, as my teacher showed us, shortening a pendulum's length makes its frequency higher. If you grab hold of a pendulum's string halfway up, so that the top segment can't move, you'll see that the frequency of the bob's swing will immediately increase.

1a. There's a related way to get a higher frequency that's commonly used on guitars and other string instruments, and also in wind instruments, especially brass instruments such as French horns and trumpets. This has to do with the concept of "harmonics," also called "overtones," which we will encounter only briefly in this book. It involves tricking a string or air column into vibrating faster than it normally does. There are only so many ways to do this, though, not enough to make complex music all by itself. This is why military bugles and old hunting horns, which can only produce harmonics of a single low note, play relatively simple tunes.

The subject of harmonics is wonderful and fascinating. It's pivotal in human music and plays a substantial role in quantum physics, especially in the details of atoms. But I can't allow it to distract us, even though that would be fun.

2. A guitar won't sound pleasant to most listeners unless its strings' resonant frequencies are chosen to be in the right proportions. When they're not, the guitar is said to be "out of tune." To put it back "in tune," the guitar player "tunes the strings," altering their frequencies by *tightening* or *loosening* them.

Concretely, a guitar string is attached at one end to a pin and wrapped at the other end around a knob. Turning the knob tightens or loosens the string, making its resonant frequency higher or lower. (If you constructed a vibrating string of your own, as I suggested at the beginning of this chapter, you can verify this yourself by adjusting how taut the string is.) Most other string instruments, including pianos, work the same way.

There's no simple analogue for this in a pendulum or a wind instrument. But a ball fastened to the end of a spring, as in Fig. 21, will bounce, showing all the features of resonant vibration that we've just encountered. You can make the spring stiffer by grabbing hold of the spring's upper coils; the coils that are still able to move will do so less freely, causing the ball to bounce with a higher frequency.[8]

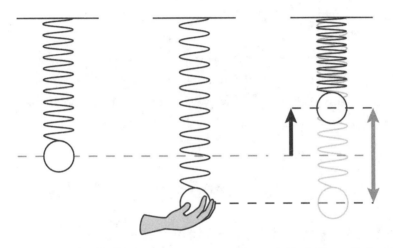

Figure 21: (Left) A ball hanging from a spring. (Center) The ball is pulled and then released. (Right) The ball bounces with the spring's resonant frequency, even while its amplitude (black arrow) gradually decreases due to dissipation. The frequency of the bounce will increase if the spring is tightened.

3. It seems a shame that guitars and pianos have to be tuned again and again. Why can't we just tune them once and then play them forever?

Making music with the instrument takes a toll, but even if you leave it in storage for a few weeks, it won't be in tune when you pull it out. This is due to changes *in the environment*. Shifts mainly in temperature and in humidity can affect the strings (or rather, they affect the guitar as a whole, with the strings adjusting along the way). More generally, most musical instruments are sensitive to their surroundings. A cold clarinet has a lower resonant frequency than a warm one; as a teen, I often found this a problem when playing on chilly mornings. Replacing ordinary air with another gas would similarly affect the frequencies of wind instruments. You've perhaps heard the high, thin voice of a person who has inhaled some helium as a party trick.

A pendulum's environment includes the gravity in which it hangs. A pendulum's frequency would be lower on the Moon than on Earth because lunar gravity is weaker than terrestrial gravity. In deep space, the pendulum's frequency would fall to zero; the bob would be essentially weightless and would float without swinging.[9]

Not all these mechanisms for changing the resonant frequencies of a guitar work on all instruments. You can't shorten piano strings while you play the instrument; that's why pianos, unlike guitars, need a separate string for each note that a pianist wants to play. A bell can't be either shortened or tightened, though it will respond to the environment, with its frequency reacting to shifts in temperature.

Despite what I've said about the universe being like a guitar, you may be wondering how it can have resonant frequencies at all. Larger instruments often have lower frequencies than smaller ones. Violins wail high, while cellos sing low. A trumpet sails above a tuba. Since the universe is beyond immense, you might imagine that its resonant frequencies should be absurdly low.

But as we'll see later in this book, the universe has a trick up its sleeve, allowing it to have ultrahigh frequencies after all. There are fields that, despite spanning the universe, can nevertheless vibrate a billion trillion times per second or even more.

Still, it's not obvious that any of the mechanisms for changing a guitar's frequencies could work for the cosmos. While the guitar's strings have ends, the universe's fields do not. They are present everywhere and cannot be shortened.[10] Since they lack edges, one cannot hope to tighten them with a mechanism analogous to the knobs on a guitar. Finally, since the universe is, by definition, everything, how could it have an environment?

Here's the surprise: *one field can serve as the environment for another.* In fact, as we will see, that's the main role of the Higgs field: to serve as an environment for a host of other fields. When switched on, it shifts their resonant frequencies, somewhat as a change in temperature changes the resonant frequencies of a guitar's strings.

This feature makes the universe more complex than a guitar. The effect of the Higgs field on other fields is somewhat analogous to allowing an aspect of one guitar string to retune the other strings. That's something human instrument makers would naturally avoid. Strings on most instruments are separated and kept independent so that any one of them can be played while having limited impact on the others.[11] But that's simply not true of the universe's fields.

══ 11 ══

The Waves of Knowing

Hearing a guitar involves a chain of events, as emphasized in Fig. 18 (p. 132). But within that chain, there's something potentially confusing. As a friend of mine put it, "How does the sound know to move outward from the instrument toward me? Is it some sort of wind that pushes the air along?"

These are natural questions with a surprising answer. When an instrument plays, *the sound waves travel, but the air does not.* This is a general feature of waves and a key to understanding the universe.

First things first, though. I'll soon try to convince you that everything around you is made from waves. We had better agree, then, on the definition of the word *wave*, as it is yet another false friend. (You just can't trust anyone in this business.)

In ordinary English, a wave at a beach consists of a single crest in the water, separated from the previous wave by a trough in front and from the next wave by a trough behind. That's what's meant when speaking of a "breaking wave" or of "catching a wave" in surfing.

But in the dialect of physics, a simple wave is generally a *series* of crests and troughs, not just one crest. For physicists, and for most scientists and engineers—including recording engineers—what we mean by "a wave" is what in English we would call a wave train, or a wave set. This is illustrated in Fig. 22: in English, a wave is a single wiggle, but in physics dialect, this whole set of wiggles is a wave.

140

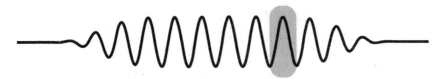

Figure 22: A simple wave in physics dialect is a chain of crests and troughs of comparable size, while in English, a wave is just one crest (shaded).

You might well ask me whether I view the water's regular upheaval at an ocean beach as due to *waves* or *a wave*? I'm somewhat torn, since I speak multiple dialects. In any case, in this book, we will use the singular form: a wave.

Similarly, as I speak, I'm making a sound wave, singular. It's a complicated, highly irregular one. That's in contrast to the sound wave that I make when I'm singing a single note or when I strike a piano key. The wave from a musical tone is much more regular than speech, a simpler vibrating shape. Though all sorts of waves, regular and irregular, are found in the universe, we will focus mainly on these simple, quasi-musical waves. An example is shown in Fig. 22; it consists of a series of crests and troughs, all of about the same height and depth, equally separated from each other.

Like a vibration, a simple wave is characterized by its frequency and its amplitude. If you've been at a beach and watched the water, not where the wave crests break on the shore but much farther out, you may perhaps have seen a bird sitting and bobbing up and down as the water undulates beneath it. The frequency with which the bird rises and falls is the wave's frequency, while the amplitude is how far the bird rises and how far it falls.

Similarly, at the end of a long dock, the frequency of the water's repeated ascent up the dock's support poles is the wave's frequency, while the height of its rise (and depth of its fall) is its amplitude. Analogous features, though with varying details, characterize many other simple waves, including sound, light, and seismic waves.

Let's return to the bobbing bird. If you've watched one, you've perhaps noticed something interesting, which bothered one of my students. "The waves are all... I mean, the wave crests are approaching the shore," she said.

"But the bird isn't being carried with them. It's just going up and down. Why isn't it brought along with the wave?"

I had the same question when I was a child. I was out fishing with my father on a little open boat, sitting with the motor off near a small pier. Wave crests were moving past us toward the dock. But I noticed that the distance between the boat and the dock wasn't changing. It was as if the wave bypassed the boat, gliding underneath it rather than pushing it along. It was many years before I understood why.

The crests of a water wave may all advance toward the shore, but nevertheless, the water isn't moving along with the crests. If it were, then as more and more wave crests reached the beach, the inflowing water would cause an ever-worsening flood.

Instead, except right near the beach where the wave crests break and the motion is complicated, the water's not doing much at all. It's just rocking a little bit, moving in small circles as the wave crests travel by. There's no overall flow that could transport the bird or a boat along with the crests. You can feel this yourself if you stand or tread water out beyond the breakers; you won't be carried beachward by the passing wave crests. This is also why you cannot surf a wave crest that isn't breaking; it won't take you with it.

The same is true for sound. The crests and troughs of a sound wave move inexorably outward from a guitar, but all the air does, in any one location, is rock back and forth a little. Earthquake waves, too, can travel hundreds of miles, but they don't carry your house with them; they just shake it. So important is this feature of waves for the workings of the universe, and for our own experience of it, that it's well worth exploring further.

The various waves I've just discussed are all examples of *traveling waves*, one of two important types of simple waves that matter for a guitar and for the universe. A traveling wave's crests and troughs move steadily, following one another in a common direction away from where they were produced. Such a wave, generated on the ocean by a distant storm, may arrive days later at your sunny beach. The sound created by a guitar; the ripples made by a stone dropped in a pond; the shaking caused by an earthquake on a fault line miles away—all are traveling waves. The most famous of the universe's traveling waves is light itself.

You can easily make such a wave. Take a very long piece of string, perhaps twenty feet long or more. (You can also use a shorter piece of thicker rope or a Slinky.) Have a friend hold one end, or tie it to a wall or a heavy chair, then pull it fairly taut. Next, wave your hand rapidly up and down a few times, then hold it still. You will see a simple wave with several crests and troughs (known in English as "several waves") move quickly down the string, as in Fig. 23.

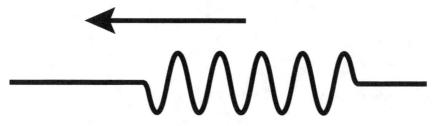

Figure 23: A traveling wave with five crests, such as can be made by wiggling a long string's end five times. It has an amplitude (its crests' height and its troughs' depth) and a frequency (how often a spot on the rope rises and falls as the wave passes).

(Because your string's length will be short, there's a distracting complication that I have to point out. The traveling wave crests will move down the string until they reach its end. Then they will bounce off the end, like a ball bouncing off a wall, and will come back toward you. Such a bounce is like an echo for sound waves. But such echoes require a nearby edge or wall, and so they are irrelevant for waves in our vast cosmos. To avoid getting bogged down in this distraction, let's try to imagine that your string is infinitely long, so that the wave's crests can travel on forever.)

Now, here's the interesting point, and the answer to my friend's original question. Even though the wave travels along the string from one end to the other, the string as a whole goes nowhere. In fact—and you can check this by putting a paper clip on the string and watching it, as in Fig. 24—as the crests of the wave progress horizontally down the rope, each little piece of string only moves vertically up and down, and not very far, either. The crests travel; the string does not.

You can also make a traveling wave in a quiet pond or pool. Lightly disturb the water rhythmically a few times. With good lighting, you'll easily

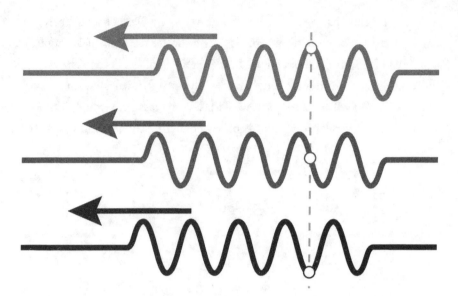

Figure 24: A wave traveling to the left, depicted at three successive moments. Though the wave travels horizontally, the rope does not; each part of the rope, such as the white dot, moves purely vertically. The dashed line shows that the dot's horizontal position does not change over time.

see ripples move outward from the disturbance. If a small object floats in the water, well away from you and from the water's edge, the ripples will rush right past the float, moving it up and down but not sideways.

At sports stadiums, when a crowd "does the wave," a wave crest of standing people moves across the stadium of seated spectators. Around and around the stadium this crest may go, faster than any person could run. But no one needs to run or even walk; each person just stands up and sits down, and no one changes seats. When the wave eventually dies out, perhaps having circled the stadium several times, everyone is where they started. The wave crest has moved horizontally with great speed, all as a result of the slower, purely vertical motion of individuals.

Traveling waves' speeds may or may not depend on their frequency. Sound waves have a fixed speed, the same for all frequencies, which assures that the multiple notes of a musical chord all arrive at our ears at the same time. But waves in water are more complicated; their speed depends on the frequency of the wave and on the depth of the water. As for waves in the universe, those of light have a definite speed that is the same for all frequencies, but this is not a general feature; for instance, waves in the Higgs field with different frequencies travel at different speeds. We will return to the Higgs field's waves later in the book.

Then there are waves that do not travel at all, which are called *standing waves*. In contrast to those of a traveling wave, a standing wave's crests and troughs are fixed in position; each crest shrinks vertically and becomes a trough, after which the process reverses and it turns back into a crest. The whole wave vibrates in place, without any horizontal motion.

Making a standing wave can be tricky at first. Hold one end of your string, rope, or Slinky—you may find it easier with a somewhat shorter length than is convenient for traveling waves—and try moving your hand rhythmically up and down, more gently this time but without stopping. If you do this at a randomly selected frequency, the string will jerk around in an irregular way. But at certain special frequencies, the string will take on a remarkable shape, such as one of those depicted in Fig. 25, with a small number of equally spaced troughs and crests. The string's motion will be cyclic: during every half cycle, each crest will become a trough, and vice versa.

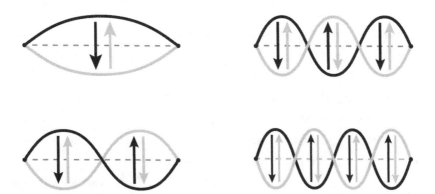

Figure 25: (Top left) The basic standing wave commonly seen on a guitar string, shown initially (black) and after a half cycle (gray). The other panels display its first three harmonics.

The more crests and troughs a standing wave has, the higher its frequency. For string instruments, all of these waves are interesting. Those with two or more crests or troughs are a string's *harmonics*. But for this book, the most important standing wave is the one with the lowest frequency, the one that most resembles a simple vibrating guitar string, shown at upper left in Fig. 25. This wave consists of nothing but a single crest that becomes a single trough and then a single crest again.

You might wonder whether this even counts as a wave; after all, I did say that a wave, for physicists, is something with multiple crests and troughs. But it turns out that standing waves can have any number of crests and troughs—even just one.

This standing wave is what is found on the string of a plucked guitar, as drawn in Fig. 20 (p. 135)—similar waves are found in many other musical instruments—and so it has everything to do with resonance. The whole string rocks from side to side, much as a child's swing rocks back and forth, with all parts of the string always moving together in the same direction. Because of its completely coordinated motion, unique among the waves in Fig. 23 and Fig. 25, this is the wave that vibrates with the string's resonant frequency.[1]

As a result of its similarity to a pendulum or swing, this standing wave is the one you'll most easily produce when you disturb a guitar string in an arbitrary way. It's also the easiest to make on a string, rope, or Slinky. Just as a swing tells you when to push, something analogous is true here. Let the string tell you how often to lift and lower your hand, and before you know it, you'll have made this standing wave.[2]

Standing and traveling waves have much in common; for instance, no matter what the crests and troughs do, any particular point on a waving string just goes up and down. This must be true, because no matter how long the waving goes on, the string doesn't go anywhere.

Yet there are essential differences between standing waves and traveling waves. Traveling waves aren't related to resonance, so you can choose and control a traveling wave's frequency as well as its amplitude. Go back to the string or pond that you were using to make traveling waves and repeat the process, but now with a higher or lower frequency. The result is qualitatively the same except that the wave's crests are closer together or farther apart.

Because a standing wave is based in resonance, it is far less forgiving. You can create the simplest standing wave on a string only by moving your hand up and down at the string's resonant frequency. If you increase the vibrational frequency of your hand, the string's simple shape will be lost; if you decrease it, the string will vibrate only grudgingly.

The idiosyncrasies of both types of waves are essential for music. As I just emphasized, a standing wave forms on a guitar string at a unique and predictable frequency. But for any sound to be heard, a traveling wave must be created in the air with the same frequency as that of the string; it's only when that sound wave reaches listeners that it can make their eardrums vibrate, again with the same frequency. For this to happen reliably, it had better be that traveling waves can be created at *any* frequency. If traveling waves required particular frequencies just as standing waves do, then most instruments' notes wouldn't be transported through the air and would never be heard. It's only because traveling waves are so flexible that all sounds made by any instrument, at any frequency, can always reach their audience.

A second important difference is that ordinary standing waves typically form in an object that has ends. You can make a standing wave on a finite string, but if you make a longer string from the same material, the resonance frequency will be lower. The standing waves on a continent-long string would vibrate too slowly for you to notice them. But traveling waves can exist on a huge string because they don't care about its resonance frequency. Similarly, sound waves exist at all audible frequencies.

You might then think that the universe would have traveling waves but not standing waves; it's just too big. But the trick I mentioned a short time ago, the one that allows the universe to have incredibly high frequencies despite its size, also allows some of its fields to have standing waves as well as traveling waves. This enhances the resemblance of the universe and its fields to a guitar and its strings; despite their totally different shapes and sizes, both can have standing waves with resonant frequencies.

Waves are found everywhere you look, and even where you don't. In addition to solids, liquids, and gases such as rock, water, and air, they appear in the ionized plasma that makes up the Sun, in traffic patterns on crowded roadways, and in large crowds of people.

In all of these cases, a simple wave involves an organized, repetitive disturbance in an ordinary substance, which is called the wave's *medium*. Air is the medium for the sounds we hear in daily life; water is the medium for ocean waves; the crowd is the medium for the wave that circles the stadium. We've seen that strings and ropes can act as *media* (plural of *medium*), too.

Like oceans and atmospheres, a typical medium fills or spans a substantial region and exists for a long time. A wave, on the other hand, is usually a transient phenomenon; temporary and fleeting. It may travel from one part of the medium to another, entering new territories and then leaving them behind. As it does so, it may lose amplitude through dissipation and eventually die out.

Though a wave is, in a sense, made from its medium, it has an almost independent existence. The medium can do one thing while a wave in the medium does something completely different. Even when there's a breeze causing the air to flow east to west, the sound waves from a guitar go off in all directions, so you can hear the guitar no matter where you sit. Once I was on a boat floating down a big river and looked on as the wave crests on its surface moved *upriver*, opposite to the current. As a crowd wave travels horizontally around a stadium, the individual spectators move vertically and don't change seats. Watching what the wave is doing doesn't tell you what the medium is doing, and vice versa.

Nor does watching the waves tell you what the medium is made from. Just because you hear sound doesn't tell you whether the gas that's wiggling your eardrums is ordinary air, pure oxygen, or pure helium. Rock can undulate whether it's granite or sandstone. Vinegar oceans and alcohol oceans would have surface waves just as water oceans do. Waves partly transcend the details of their medium.

Strikingly, waves show profound similarities to ordinary objects despite their obvious differences. Imagine (or try) this do-it-yourself experiment, illustrated in Fig. 26. Put a little table outdoors and place a small cup on it. Now, take a rubber ball and throw it gently at the cup. If your aim is good and the ball hits the cup, chances are that the cup will be knocked over and will tumble off the table into the grass. The ball, too, will end up in the grass, bearing witness to its role.

Put the cup back on the table, and stretch a long rope across the table so that it sits quite close to the cup but doesn't touch it. The rope should be fairly tight, but not overly so. Now, create a traveling wave on the rope. If you do it right, the wave crests will move down the rope and strike the cup, knocking it over. Soon after, the wave will dissipate, leaving the rope in its original condition.

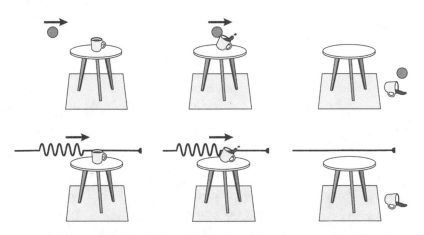

Figure 26: (Top) A ball can be thrown to knock a cup off a table. (Bottom) A wave on a rope that passes near the cup can do the job, too.

Thus, a ball, given some motion energy, can travel through empty space and transfer some energy to a cup, knocking it over. And a wave, given some energy, can traverse a rope (or some other medium) and transfer some of that energy to the cup, knocking it over.

The ocean doesn't go far in a storm, but the waves created by its winds can erode beaches and smash our boats and homes to pieces. The rock underneath us doesn't go anywhere, but the ripples from a big earthquake can knock down distant bridges and apartment houses. The atmospheric waves from a powerful explosion can shatter windows miles away. Energy can be used to make waves in one place, and the waves can travel great distances and transport that energy to an entirely new place, where it can be useful or destructive.

Balls and waves can both carry information, too. Need to tell a teenager that it's time to eat? You could toss a balled-up piece of paper, with the word *dinner* written on it, onto your teen's computer keyboard. Or you could just send sound waves by yelling, "Dinner!" Good luck; the two methods are equally ineffective.

Despite these parallels, there are differences worth noting. When you started with a ball in front of the table, you ended up with a ball behind the table. Ball and cup are both in new positions. In the other case, you started with a rope stretched across the table, and the rope ended up right where it started. Only the cup moved to a new place.

A ball is obviously a physical thing, a material object with substance and heft, something you can hold in your hand, toss and catch, cut in half, weigh on a scale, and balance on your nose. More generally, the ball exists on its own and can go anywhere you choose to throw it, whether north, south, east, or west. We can't do any of these things with a wave on a rope, which is fleeting, ephemeral, and trapped within its medium.

So even though it's intriguing that the wave and the ball can have such similar effects, perhaps we shouldn't read too much into it. Our intuition tells us that at heart, they're fundamentally different. After all, though a wave involves a vibration *of* something material, it itself is not material.

Or ... *is it?*

Soon, when we confront the quantum nature of the world, this question will be turned on its head.

12

What Ears Can't Hear and Eyes Can't See

One of the great pleasures in life, for me anyway, is experiencing a sunset over a lake or an ocean. What a feast for the eyes and ears! Clouds glowing orange, rose, and luminous gray, their reflections lighting the water; birds calling to one another amid the gentle lapping of waves on the shore; the rustle of wind in the trees as the sky turns ever deeper shades of blue. An evening like this makes me appreciate our senses.

Yet our senses are far more limited than common sense might suggest. We are, after all, largely deaf and almost entirely blind.

We are so accustomed to our perceptions that we rarely consider how they actually occur. Expressions such as "I hear a guitar" or "I see a guitar" are abbreviations that obscure intricate, complex processes, much as the word *elementary* obscures what should really be "up-to-now-apparently-elementary." Usually, such linguistic shorthand is harmless. But if we are aiming to understand the universe at its most fundamental level, then what is hidden by speech, thought, and perception becomes important.

The vibration of a plucked guitar string creates a sound wave that travels across the room and encounters your eardrums. Only then does the process of hearing begin. The sound wave makes your eardrums vibrate, creating waves in your cochlear fluid that are detected by tiny hairlike structures called stereocilia. From there, electrical signals are sent down your auditory nerves to your brain, which processes the signals and somehow gives you the conscious experience of a musical tone.

In this process, your ears and brain never directly engage with the guitar. They engage only with the sound waves that have entered your ear canal; those waves are the only things you actually *hear*. Similarly, all you *see* is the light that has reflected off the guitar and reached your eyes. We rely on traveling sound and light waves to bring us information, on our eyes and ears to detect those waves, and on our brains to make meaning out of them. But we don't hear or see the objects that create or reflect the sound and light; our brains merely infer their existence. Our knowledge of them is entirely indirect.

All of our sensory organs take in information only when it reaches our bodies, not before. What they learn is then used by our brains to gain some idea of the objects around us and to create for our consciousness a picture of the outer world. We experience that picture as though it were reality, unaware or forgetful of the fact that it is a partial reconstruction of the outside world and in no sense a direct image of it. Everything we know of the environment around us is both indirect and incomplete.

One afternoon during a college class, my students complained of an electronic whine in the room. I heard nothing. This is no surprise; young people can hear high-pitched sounds that middle-aged ears can no longer detect. Still, even children are deaf at higher frequencies; that's the principle behind a dog whistle, inaudible to humans but easily perceived by our canine friends. *Ultrasound*, famously used to destroy kidney stones or create an image of a fetus in the womb, involves sound waves too high-pitched even for dogs.

The sonic world of elephants also extends beyond human hearing, but in the other direction. This was recognized only when scientists studied elephant communication with equipment sensitive to *infrasound*, sound waves whose frequencies are too low for the human ear.

It's not surprising that our ears, sophisticated as they are, have limitations. Any physical device used for measuring the world will have them. But it's noteworthy that *our brains don't warn us about these limitations.*

As a friend of mine once remarked, "It's true that as a child, I assumed my ears could hear everything and that I would only miss sounds that were too quiet. I wasn't intuitively cautious that there might be *loud* sounds that I couldn't hear."

For most of human history, the world of loud unheard sounds was unknown even to adults. Though the ancient Greeks suspected its existence, it's been explored only in recent centuries.

Then there's the unseen world, full of bright light of all kinds. We're as blind as we are deaf—more so, in fact. By this I don't mean that we are less perceptive than hawks and eagles; in the context I'm referring to, their eyesight is hardly better than ours. Nor would ordinary glasses help us. Until the dawn of the nineteenth century, there was only one hint that we're so blind. Nobody read it correctly until after the fact, even though it was plain as rain.

"What hint is that?" asked my friend.

"The rainbow," I replied. "Have you ever wondered why it's so narrow?"

He reflected for a moment. "Never did till just now."

A rainbow is not a material object, with rest mass. It's a play of light, something like a reflection in a mirror. But mere reflections of light, like echoes of sound, don't transform the light so dramatically. The key to a rainbow is that raindrops, unlike mirrors, are transparent.

Sunlight doesn't just reflect off a raindrop; it enters it. When it does so, it *refracts*, meaning that its path is bent. Then some of it *reflects* off the back side of the drop. Finally, it *refracts* again as it exits the drop, now traveling backward, but not quite the way it came in (Fig. 27). The direction of its motion is shifted by a certain angle, which means that every raindrop that lies at that angle, relative to the direction of the incoming sunlight, will send light back to your eye. If you were on a cliff or in an airplane looking both up and down into a rainstorm, you'd see a perfect circlet of light. The only reason most rainbows are arcs is that we usually look at them from the ground, peering upward into the rain. From that perspective, the lower portion of the circlet is blocked by the Earth's surface.

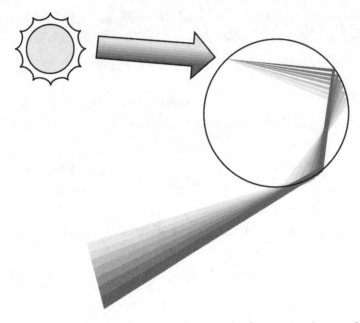

Figure 27: A refraction, a reflection, and a second refraction inside a raindrop break up sunlight into its various frequencies, some of which we can see as a range of colors, and send them out of the raindrop at slightly different angles.

You might then expect that we'd see a band of white light, as colorless as sunlight reflected off a window. However, in refraction, the paths of lower frequencies of light are bent slightly less than those of higher frequencies, and so they exit the raindrops at slightly different angles. Because different frequencies of visible light are reconstructed by our brains as different colors, the circlet of light has varying color; the outside of the band of light appears red-orange, the inside is violet-blue, and yellow-green is in the middle.[1]

That's how we end up with a pretty sight. Still, why is the rainbow so narrow?

It's not. It only looks that way. Blame the blindness of the human eye.

A rainbow is a vast, broad band extending across much of the sky. We perceive only a little slice of it; our eyes' opsin molecules absorb very little of its light, so most of it goes unseen. Above the visible rainbow is a thick swath of infrared light, and below it is a similar swath of ultraviolet light (Fig. 28).

Figure 28: A rainbow appears narrow only because we are blind to
both infrared (IR) and ultraviolet (UV) light.

Nobody knew this until the year 1800, when William Herschel, discoverer of the planet Uranus, observed invisible light above the red in the rainbow. This *infrared light* is easily absorbed by your body, causing your skin to become warmer and your heat-sensitive nerves to fire.[2]

Less than a year later, Johann Ritter looked for and found ultraviolet light below the rainbow's violet band. Light at these higher frequencies is invisible to the eye but capable of causing suntans, sunburns, and skin cancer. To prove that it exists, Ritter exploited the fact that it can also cause certain chemical reactions, such as the blackening of silver chloride.

This was the thin end of the wedge. Scientists gradually realized that visible light represents a tiny sliver of the range of possible frequencies for what we now call *electromagnetic waves*. The name reflects an intricate connection with electric and magnetic fields, which we'll encounter in coming chapters.

Our experiences of sound and of light feel dissimilar. But this is a feature of our inner worlds, arising from physiological differences between our auditory and visual systems. Out in the real world, simple sound waves and light waves have a lot in common. Each has just three properties: frequency, amplitude, and speed.[3] Traveling at the speed of sound, a sound wave has a frequency that determines whether it's audible and what pitch we experience, while its amplitude controls its volume. Analogously, as a light wave cruises at the cosmic speed limit, its frequency determines whether it's visible and what color we experience, while its amplitude controls its brightness.

Not only do our brains provide no hint of this similarity between light and sound, but they also conceal that we are far more blind than we are deaf. Humans can hear sound waves with frequencies as low as twenty cycles per second and as high as twenty *thousand* cycles per second. In musical terms, that range of one thousand from lowest to highest frequencies covers roughly *ten octaves* of sound waves. You can find over seven of those octaves on a piano keyboard.[4]

But when it comes to light, our eyes are limited to frequencies between about 430 trillion and about 790 trillion cycles per second. That's a range of only 790/430, about 1.8, which is much smaller than our hearing range; in musical terms, *it's not even a single octave!* For context, the full range of electromagnetic waves, called the electromagnetic spectrum and illustrated schematically in Fig. 29, is known experimentally to span well over 150 octaves, with frequencies ranging from below one cycle every billion years to above a billion billion billion cycles per second. It likely goes much further upward. We can barely see any of it.

Because the electromagnetic spectrum is so broad, scientists found it useful to divide it into sections, as shown in Fig. 29. These completely arbitrary categories include radio waves at the lowest frequencies and *gamma rays* at the highest; in between are found microwaves, infrared light, visible light, ultraviolet light, and X-rays. Remember that these are just arbitrary names for frequency ranges selected by Euro-American scientists. Nature makes no such divisions or categorization.

Let me remind you of something I mentioned earlier, whose importance makes it worth emphasizing again. Even though, for humans, waves of visible light seem dramatically different from waves of invisible forms of light, there's nothing intrinsically different about them except for how often they vibrate. All the differences are inside our heads, literally: the opsin molecules in our eyes absorb only photons that lie in the visible range of frequencies. Light (and its photons) at other frequencies elicit no response.

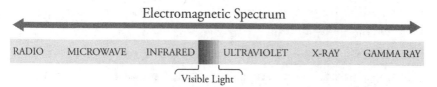

Figure 29: The spectrum of electromagnetic waves, illustrating schematically the continuous range of frequencies from lowest at left to highest at right. Also shown are the arbitrary divisions by scientists of the invisible frequencies into regions, and the very narrow range (not drawn to scale) that is visible to the human eye.

"Now, why is it," a friend asked as we sat outside on a summer evening, "that the rainbow only contains some colors and not others? I mean, there's no silver or pink or brown in a rainbow, is there?"

"No, there isn't," I agreed, "and the fundamental reason is that although frequency is part of the physical world, color is part of the biological and psychological world of a human being, and the two don't line up."

"You mean that frequency is what exists in the outside world, color is how your brain perceives it, and the perception doesn't match the reality?" she asked.

"That's right. With sound, it's pretty easy: if a musician plays three different notes at the same time on a guitar or piano, you'll experience all three of them simultaneously. It's what we call a *chord*.

"But if I flood your eyes with three different light frequencies," I continued, "your eyes will take that information, throw away most of it, and scramble the rest. Then they'll send what's left on to your brain, which scrambles it even more. And by the time it's done, you'll have the conscious experience of *one* color—usually a color that is absent from a rainbow. Moreover, that psychological experience may not have much or anything to do with the psychological experience you'd have if you saw each of those frequencies separately."

"Oh!" her husband interjected. "Is this related to why red, green, and blue pixels on a video screen are used together to make white?"

"That's a famous example," I replied. "And what it tells you is that white light doesn't exist in the physical world. What exists are *combinations of electromagnetic waves, with different frequencies, that the human eyes and brain perceive as white.* White is an experience, a psychological state of the human mind, not a physical phenomenon in the outside world.

"For instance, as you said, suppose you take light waves from the rainbow that are red, green, and blue. Then you send them into people's eyes from the same direction at the same time. What they will experience is the color *white*, which couldn't be more different from the psychological experiences of *red*, *green*, or *blue* separately or even side by side.

"But that's just the beginning! If you take a completely different combination, perhaps ten light waves with various frequencies spread across the rainbow in just the right way, humans will again experience *white*. There will be no hint that the underlying pattern of light is any different from the red-green-blue combination that is also perceived as white. In fact, there are an infinite number of ways to create the experience of white light; for example, the white of sunlight is literally made of all the colors in the rainbow and thus is completely different from the simple red-green-blue combination used by a video screen.

"The same is true of colors like pink and brown and magenta; each of these color experiences can be created by combining light waves of different frequencies in all sorts of different ways. And in none of these cases does your brain tell you what you're actually seeing."

"That's amazing!" my friend exclaimed, sitting back in her chair. "So for some colors, like blue, we see the world as it is, whereas for other colors, like pink, we're seeing a mishmash of reality?"

"Not even blue is exempt," I replied. "It's true that if I shine a simple electromagnetic wave with a definite frequency from the blue part of the rainbow—say, a wave with a frequency of 650 trillion cycles per second—into your eye, you'll experience blue. But the sky appears blue, too. And it isn't."

"Did you just say *the sky isn't blue?*" her husband almost shouted. This generated some stares from nearby tables.

"The Sky Is Not Blue," I declared in a subdued but emphatic tone.

"What is it, then?" he asked impatiently.

The sky just appears blue to the human brain. In reality, the visible light coming from the sky is a rich blend of waves with all the frequencies found in the rainbow. The waves from the bluish band of the rainbow have somewhat larger amplitude than do those in the greenish and yellowish bands, which in turn exceed those at orange and red frequencies. But as your eyes absorb and process the sky's light, they pare away its complex details, reducing them to a very small amount of information. Only that diminished information is sent to your brain, which then processes it further and creates an experience of *sky blue* in your consciousness.

In short, not only can't our eyes perceive most frequencies of light, they're not even able to capture most of the details of *visible* light! Our eyes are as much censors, with a *c*, as they are sensors, with an *s*. They're unresponsive to everything except the visible frequencies, and even the information about visible frequencies is drastically edited before it is sent to our brains for interpretation. Then our brains are like a government information agency in a totalitarian state: they take the already censored information and transform it into the world they think we ought to know.

The world we think we see isn't the world as it is. At best, there's a resemblance. Our eyes aren't transparent windows that show us things as they are; instead, our visual systems *create images* for us, much as cameras and their attached video screens do. Other animals, with slightly different eyes and brains, can't possibly experience the world the same way we do; the correspondence between frequency and the color in their consciousness is surely different. Even some humans have unusual visual systems with stronger or weaker color perception than the majority of us; I'm sure their experiences aren't like mine. And it's impossible to imagine what the world would look like to the consciousness of an alien species, with light detectors of its own.

Color is both much less than and much more than frequency. This is a rich and complicated story, with entire bookshelves in libraries dedicated to it. For us, the important lesson is simple: human experience does not passively and faithfully reflect the universe around us. What we experience is heavily censored, processed, and reconstructed.

My friend shook her head in wonder. "In politics, nothing is as it seems," she mused, "but I didn't realize the extent to which it's true in basic perception."

She thought for a minute, looking up at the deepening "blue" of twilight, where the first stars were starting to appear. Then she added, "I suppose evolution only has to ensure that human experience is rich enough, and relevant enough, that we can survive and have kids and maintain the species. Since that's all that's necessary, why should we expect our conscious experiences would correctly capture reality?"

I nodded. "We're optimized for survival, not accuracy. And with our limited brain power, there's no way that we can handle the flood of information from the outside world; if our senses didn't censor what's gushing in, we'd be completely overloaded. But it does mean that we grow up with a very narrow conception of the physical universe. It's only through scientific instruments that we've come to realize how much we were all missing."

The last two centuries have provided us with numerous technologies that enhance our senses beyond their biological capability. Infrared goggles, for instance, extend the frequency range that we can see by electronically converting invisible infrared light waves into waves of visible light. This is particularly helpful at night, when living creatures are significantly warmer than their surroundings and, though invisible to our eyes, glow more brightly in infrared light waves than do the air and ground. Many other scientific instruments, including various types of telescopes, not only magnify an object but convert an invisible phenomenon into a visible image. Nor is this limited to sight: an ordinary radio is a device for turning radio waves, a form of invisible light, into sound waves that our ears can hear, while an ultrasound scanner converts unheard sound into a visible-light image. We use sensory enhancement every day without even thinking about it, making use of features of the world that not long ago lay beyond our grasp.

People who claim to believe only what their senses tell them are missing out on the vast majority of what there is to know in the universe. They are also deceiving themselves. After all, even a cell phone exploits the unseen and unheard and unfelt. The fact that modern gadgets seem like magic, making use of the universe beyond human senses, points out yet again the weakness of common sense in the physical world.

While sunlight seen directly appears white, and sunlight scattered by the atmosphere appears blue, the situation changes after sunset. Moonlight and starlight scatter off the atmosphere, too, but the scattered light is diffuse and feeble. This explains why the majority of the sky between the stars appears black to us: when our eyes detect no light waves, our brains produce the conscious experience of "black."

Yet the poetry of a pitch-black sky is another consequence of our blindness. In fact, the universe is still filled with light left over from the blazingly hot birth of the universe nearly fourteen billion years ago. If our eyes weren't blind to microwaves, with frequencies a thousand times lower than those of visible light, we'd see the whole sky glowing faintly with what is known as the Cosmic Microwave Background (CMB). It was the surprising hissing of a microwave receiver in 1965 that tipped off Arno Penzias and Robert Wilson. After ruling out pigeon droppings and interference from human radio communication, they concluded that the night sky shines after all. To different eyes, the universe would never seem dark.

The CMB represents a minor wrinkle in this book's story. An important reason that deep space, way out yonder, is not the same as empty space is that it is filled with this tenuous bath of microwave photons. With precise scientific equipment, you could measure your rate of travel relative to the CMB and use it to get clues about your motion, even out in deep space.

Despite this, the CMB leaves Galileo's principle intact. First, a strict application of that principle requires an isolated bubble, which the CMB's photons can't enter, so you can't use them to determine your motion inside such a bubble. Second, in the more informal application of relativity to nearly isolated bubbles, such as the Earth, an airplane, or an atom, the CMB is ineffectual, as it's too diffuse to have any impact. (It does create an extremely tiny bit of drag in outer space, negligible in almost every circumstance.) And third, from the CMB, you can determine your motion only *relative to the CMB itself*, which is by no means the same as measuring your motion relative to empty space! For these reasons, I will mostly relegate the CMB to side comments and endnotes. We shouldn't forget about it, but it doesn't affect the main conceptual points of this book.[5]

The very fact that we can detect the CMB's photons, not to mention those of starlight, raises a puzzle. It's one you might not have ever noticed.

In my early twenties, I chatted on a long bus ride with a scruffy college student who was sitting next to me. He was wearing a tie-dyed shirt, and he had a guitar with him. He had a lot of physics questions.

"I know there's no sound in outer space," he said.

"Right," I agreed, "Without air or anything else, there's no sound."

Sound can't travel in *any* empty space, including an artificially created vacuum.[6] Imagine you put a guitar in a glass box, removed all the air from the box, and remotely plucked one of its strings. The string would vibrate as it always does, with a standing wave. But you'd hear no sound, because without air around the guitar, no traveling waves would be created that could carry a sign of the vibrating string to your ears.

"But even though I couldn't hear my guitar in outer space, I'd still see it, right? So if sound's a wave in air, what is light a wave in?"

Ah, yes. What is the medium for light? What supports light waves as they travel from distant stars to Earth? For fifty years, this was a central issue in physics. The question has a very strange half-answer, so bizarre that it took Einstein to figure it out.

It had been understood for centuries that you can't have sound waves without air or water waves without an ocean. Once it became clear to nineteenth-century scientists that light is a wave, it seemed obvious that it has to have a medium. This medium was called the *luminiferous aether*, a wonderful name that I find simultaneously beautiful, delicious, and demonic. This aether must be present in every part of the universe, even within ordinary objects, so that light can cross the cosmos from distant galaxies, emanate from atoms in glowing coals and candles, and even (as radio waves or X-rays) pass through walls.

Despite its lovely name, decades of attempts to find evidence of it failed. The world's greatest physicists puzzled over the issue.

Then, in 1905, young Einstein suggested that light isn't like sound after all. The luminiferous aether, he proposed, doesn't exist.

Or perhaps it does. But if so, it can't be detected, not even with seemingly foolproof methods, and its properties are so peculiar that it verges on impossible.

FIELDS

===

Though ocean waves freely roam the seas, their travels end when they reach land. Earthquake waves can cross the Earth but are confined to its rock. Sound waves stop at the edge of the atmosphere. But light waves keep going.

It's hardly surprising that ocean waves can't head into outer space; where there is no water, there can be no water waves. What's surprising is that light waves don't appear to face a similar impediment. They seem to be able to go *everywhere*.

This suggests that light's medium must be an everywhere-medium, found throughout the universe. But if that's true, in what sense is empty space ever actually empty?

Particles love empty space; they can effortlessly coast across it. Waves fear it, as they are condemned to remain within their medium. Conversely, waves can steadily travel across their medium, while for particles, a medium is an obstruction that typically drags on them and slows them down. Given these incompatible preferences, it's perplexing to imagine both particles and waves coasting freely across our seemingly empty cosmos.

Long before photons were discovered, many scientists suspected that light is made of particles. This is because visible light seems to travel in straight lines instead of bending around corners, as sound and ocean waves do. Among these scientists was Newton. He had assumed that outer space is truly empty when he freely applied the coasting law to his understanding of planetary orbits. This assumption seemed compatible with light made from particles but not with light made of waves, which would find truly empty space an impenetrable barrier.

Yet the possibility that light is a wave had already been proposed, not long before Newton, by the Italian physicist Francesco Grimaldi, based on his own experiments. He was soon joined by Huygens in this view. The competing hypotheses were debated until 1802, when wavelike properties of visible light were demonstrated beyond doubt by Thomas Young. Newton's guess was wrong.

This made the paradox sharp. Since light is a wave that can travel from the Sun and the stars to Earth, its medium, the luminiferous aether, apparently fills the universe. Empty space, then, isn't entirely empty. If that's true, though, why didn't Newton have to account for drag from this aether when calculating the paths of the planets and the Moon? Why haven't billions of years of friction slowed the Earth, or at least stripped away its atmosphere? Why does it seem as though everything we and the Earth are made of— electrons, protons, neutrons—can coast unimpeded through outer space, much as light waves do?

We might briefly wonder if these questions are missing the point. Perhaps Earth and other ordinary objects evade the problem. For instance, we might imagine that they wear a protective coating, something that allows them to plow safely across the cosmos while keeping the luminiferous aether at bay. But then we remember how the Earth and Sun, the walls of a spacecraft, a human body, and anything else we usually encounter are themselves mostly empty space. Even solid rock is no more than gossamer dust. So as we travel the cosmos, it's not just that our planet and our bodies go through what seems like empty space. To the same degree, *empty space goes through us.*

Clearly this transgression of our bodies is unproblematic. We don't even notice it.

Because of this, the question of whether empty space is truly empty is relevant to you and me, as is the puzzle of how both the atoms we're made of and the light waves we see can cross apparently empty space. Whatever media might be found in empty space, they are present everywhere and always. We coast through them, throughout our lives, as though they're not even there. Somehow, unlike familiar media such as water or air, they let us and everything else fly through them at tremendous speeds without resistance, thereby maintaining Galileo's principle and the coasting law. So if there really are media in seemingly empty space, these substances must be extremely thin, exceptionally clever, or both. If instead they are absent…well, then we're back where we started; how does starlight cross the emptiness and reach the Earth?

Into this bubbling cauldron of confusion we may stir another ingredient: the forces that affect objects separated by empty space. Not only gravity but also electric and magnetic forces allow distant objects to pull on one another; for instance, Earth's magnetism steers electrons from the Sun to our planet's poles, where they generate the northern and southern lights. Is the cosmic reach of these forces, out across seemingly empty space, made possible by the *presence* of some sort of medium or by the *absence* of any such medium?

These conundrums and paradoxes are profound and troubling. The concept of *fields* will help us address them. Along the way, we'll be forced to grapple with the extraordinary properties of empty space and to recognize it as a sort of medium, though one unlike any we are familiar with. Our confusion will abate somewhat, enough for us to proceed with our initial concerns: the mysteries of motion and the origin of the energy that brings about rest mass. But don't expect comprehensive, satisfying resolutions to all these puzzles. I don't have them. No one does.

With that, we come to the most difficult section of this book. It was challenging to write, as I knew it would be, and it is challenging to read. It introduces concepts that are unfamiliar, eerie, and slippery even for physicists. Already in these opening pages, I've confronted you with paradoxes—particles versus waves, empty space versus a medium—whose resolution is as mysterious as the paradoxes. Experiments confirm that what I'm about to tell you is a correct story, but it seems unlikely to be the complete story. The nature of empty space and its fields remains a subject of active scientific research and debate.

In this book's opening pages, I suggested that you might find it helpful to read certain sections of this book more than once. If there were any such section, this would be it; or you might consider revisiting this section and the following Quantum section together after a first pass. One thing I'm sure of: before I became an expert myself, I would have needed to read this material twice. The subject is full of strangeness, and if you have trouble making sense of it, remember that physicists do, too.

13

Ordinary Fields

Not long after the Higgs boson's discovery, I was sitting outdoors at a cafe, explaining the significance of the Higgs field to an old friend. When he asked me what a field is, I pulled my keys out of my pocket, threw them into the air, and caught them as they came back down.

"So how did the Earth do that?" I asked him.

"How...*what*?"

"Somehow the Earth grabbed hold of my keys and kept them from flying away. How did it manage to do that without touching them?"

He gave me a confused look. "Gravity, right? Am I missing something?"

I grinned at him. "Yeah, it's gravity, sure," I agreed. "But isn't there something strange about it? It's kind of a magic trick."

My friend accused me of pulling his leg. I wasn't, but something else was.

"The Earth doesn't pull directly on the keys. Instead, it relies on an intermediary—a third party—known as the gravitational field. Unlike our planet, this field exists everywhere, in and around the Earth and all across the universe. It's the field that crosses the gap between the ground and the keys."

The gravitational field isn't the only one that can act as an intermediary across a gap. You probably have magnets somewhere in your house, perhaps attached to the front of your refrigerator. I have a bunch in a drawer, all stuck together. But the stickiness of magnets is not like that of sticky tape, which is sticky only when you touch it. Instead, if you separate two magnets and hold them close, you can still feel the pull between them. Putting a sheet of paper or cardboard between them doesn't eliminate this pull. There's an unseen force at work, trying to bring them together even though they're not in contact. The responsible party is the magnetic field.

The same is true of socks just out of the dryer. They may be stuck together, but if you separate them slightly, they'll still attract one another. Run a comb through your hair a few times, then bring it close to your head; your hair will rise toward the comb. In these examples, the intermediary is the electric field.

Unseen forces across gaps are common in nature, allowing objects that aren't in direct contact to affect one another nevertheless (Fig. 30). This kind of "action at a distance" seems magical. But from the modern scientific perspective, it's not that different from something completely familiar.

Figure 30: (Left) The magnetic field around a magnet (black square) can rotate a compass that would normally align with the Earth's steady magnetic field. (Right) The electric field around a recently used comb can pull on tissue paper without contact between comb and paper; this is an example of "static electricity."

I place a tennis ball on my bed. I then sit down abruptly near the bed's edge, and the ball rolls toward me. The mattress has served as an intermediary; I've made it bend, and the resulting tilt has caused the ball to move. But suppose the mattress were invisible; then this sequence of cause and effect wouldn't be obvious to spectators.

I tie a long string around a shoe and, holding the string's other end, walk across the bedroom. I can now make the shoe move with a sharp tug on the string, which acts as an intermediary between my hand and the shoe. If you were watching but couldn't see the string, you might be impressed by my apparent command of magic.

A fan is pointed toward a room's center and turned on full blast. Though the windows lie to the side of the fan, their curtains rustle anyway. That's because the strong wind near the fan leads to light wind currents throughout the room. In this example, the invisible wind serves as intermediary between the fan and the curtains.

Compared to these ordinary examples, the pull between the Earth and Moon or between two magnets seems stranger, but that's only because the intermediary cannot so easily be seen or felt. This reflects a failing of the human senses, not something unreal about the intermediary. Our senses have disappointed us before, giving us no inkling of ultrasound or X-rays, so their silence can hardly be taken as evidence. These intermediaries can be easily and reliably detected by scientific instruments, and in sufficiently extreme circumstances, they *can* be felt by the human body.

We will soon see that fields can do much more than be intermediaries between objects; they are also responsible for the objects themselves. But this is for later chapters.

Let's first explore the most familiar of fields, easily felt by every human being. That field is the wind—the flow of air.

As a child, I developed a lifelong fascination with weather. Floods, tropical storms, and blizzards pounded our rural Massachusetts town in my early years, sometimes cutting us off from the electrical grid. Once a tornado dissipated scarcely a quarter mile away, its inflow stripping trees and plastering their leaves against our plate-glass window as we sat in our living room, unaware of the potential danger. Such demonstrations of natural power left me in awe, an awe that was only to grow as I became more aware of nature's arsenal.

Among several toys and tools that I acquired over those years was a device for measuring the wind. It consisted of a hollow vertical plastic tube with a horizontal hole at the bottom through which the wind could enter. Inside the tube was a tiny lightweight foam ball. On a windy day, when I pointed the hole in the direction from which the wind was coming, the ball would rise. Referring to a scale on the side of the tube, I could infer the speed of the wind from the height of the ball.

With this simple wind meter, I could measure the wind's speed and direction. But one such device can measure the wind in only one location. Imagine now a million children scattered evenly around the world, including on the oceans, each one with a little wind meter. As they all record the wind and report back to us at a central station, they allow us to assemble a detailed picture of everything the wind is doing at this moment across the Earth's surface. Indeed, this is part of what weather forecasters have to do. They even display the surface wind using maps similar to Fig. 31, which depicts the wind across the United States on a particular day. The lines' brightness indicates the wind's speed, while their orientations indicate its direction, predominantly west to east but spiraling dramatically into a storm near the East Coast.

But for weather forecasters, this is not enough. They need to know the wind throughout the atmosphere. So let us furnish the million children with

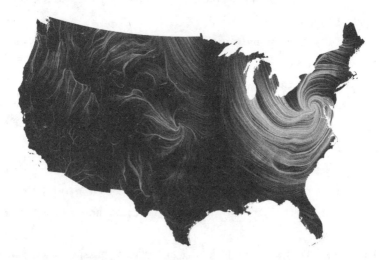

Figure 31: The wind field (with lines' orientation and brightness corresponding to wind direction and speed) just above the surface of the United States on October 30, 2012, soon after tropical storm Sandy came ashore from the Atlantic Ocean. "Wind Map" by Martin Wattenberg and Fernanda Viégas (hint.fm/wind).

dozens of helium balloons, each carrying its own wind meter. Launching the balloons according to a prearranged schedule, we can gather knowledge of the wind throughout the atmosphere's three dimensions: latitude, longitude, and altitude.

Through this elaborate operation, we can build up a complete database of the wind across Earth's atmosphere at a particular moment. This database fully describes a property of the air—the rate and direction of its flow—everywhere that air is to be found. In scientific language, we have now ascertained the *wind field* at a moment in time, capturing the full information about what the wind is doing everywhere. The wind field at any particular location is simply the wind that we would measure, there and then, with my little wind meter.

The task of scientific weather forecasting is to take this information, along with that of other fields (such as the air pressure at each location in the atmosphere), and predict what the wind field will be in the future. To do their job with confidence, forecasters need as much of this information as they can obtain. A meteorologist who measures the wind in Paris on Monday, but nothing else, can't possibly predict the wind in Paris on Thursday. At a minimum, knowledge of the wind above Europe and the Atlantic Ocean, from the surface and up several miles, is needed. It's out of the whole wind field, captured in all that information, that the future's weather will be born.

The wind field provides us with a typical example of an ordinary field: it is a changeable property of an ordinary medium (the flow of the Earth's air), and to know it fully and predict its future behavior, we must measure it everywhere. Many other ordinary media and ordinary fields appear in daily life and in physics courses, and to predict what they will do tomorrow based on what they are doing today is a classic goal of science.

Water is another medium, and one of its properties is pressure. If you know the pressure everywhere in the ocean, then, by definition, you know the ocean's pressure field. Like the wind field, the pressure field of water is not to be trifled with; it can crush a submarine that dives too deep.

The Earth's rock has a property called *mass density*—the mass of a chunk of rock divided by the chunk's volume. For example, granite rock has a mass density of about 2.5 grams of material in each cubic centimeter, while the Earth's inner core has a mass density four times higher. Mass density has all the hallmarks of a field; it is present everywhere within the Earth, captures one of its properties, and can change predictably over time.

Rock has other fields, too. Layers of sedimentary rock, such as sandstone formed by sand and mud laid down over millions of years, are sometimes seen lying perfectly horizontally in the sides of mountains or canyons. But sometimes they appear curved, showing distortion by powerful geological strain. The degree and direction of this curving of the layers is a property of the rock, which we might call its "bending field."

Here's a more subtle example. Viewing a large block of iron as a medium, let's consider its property called *magnetization*. That's the degree to which (and the direction in which) each little piece of the iron is magnetized. This might sound odd, since we're used to thinking of objects as being magnets—i.e., being magnetized as a whole—or not magnetized at all. But partial magnetization is possible, as sketched in Fig. 32, and the amount of magnetization of the iron from place to place is an ordinary field. (Be careful not to confuse the magnetization field with the familiar magnetic field; the former exists only *inside the iron*, while the latter, a property of the universe that can orient compass needles, exists *everywhere*.)

An ordinary field characterizes a property of an ordinary medium, but it may not be obvious which one. The wind field tells us about air's flow, and the bending field captures the distortion of sedimentary rock layers. But what property of iron is magnetization?

Though magnets were used and manufactured for centuries, and though the magnetization field was studied even in the 1800s, an answer to this question wasn't provided until the twentieth century. Then it was learned that in materials such as iron, the atoms themselves act as tiny magnets. In an unmagnetized piece of iron, the atoms point in random directions and their magnetic effects cancel each other out. In a magnetized piece of iron,

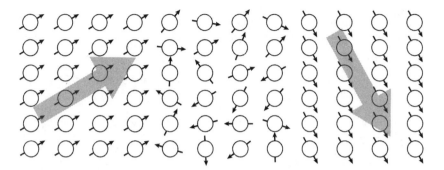

Figure 32: Atoms in an iron block; each acts as a tiny magnet (black arrows). Where they are randomly oriented, as at center, there is no net magnetization, but where they are aligned, there is a measurable magnetization field (gray arrows). Not to scale; the number of atoms is far greater than shown.

however, the atoms align, and together, they cause the iron to behave as a perceptible magnet that can stick to a refrigerator door.[1]

Our senses can't detect the magnetization field across an iron slab. But the universe knows it's there, as you can confirm by bringing a needle close to the iron's surface. If the magnetization field just beneath the needle is zero, the needle won't budge, but if the magnetization field there is strong, the needle will jump to the metal and reorient itself until it aligns with the magnetization field's direction. In fact, this trick allows us to measure the magnetization field from outside the iron, and we'll use it again soon.

We now have a variety of examples, some familiar and some less so, in which a medium has a property that we can characterize as a field. Let's now look more closely at the relationship between field and medium.

Crucially, *a field of a particular type may arise from many different media.* Wind, for example, isn't limited to Earth. There are dust storms on Earth, Mars, and Titan (Saturn's largest moon), implying that all three bodies have atmospheres with wind fields. But their atmospheres differ: Earth's is made mostly of nitrogen and oxygen, that of Titan is mostly nitrogen, and Martian air is predominantly carbon dioxide.

Do the dust storms on the red planet have the same cause as those on our own? It depends on how we think about the question. On Earth, a dust storm is caused by the rapid flow of nitrogen and oxygen, while on Mars, flowing carbon dioxide is responsible. But I hope you agree that this distinction is beside the point. In both cases, it's the wind that displaces the dust and transports it to high altitude. The fact that the wind fields of the two planets are founded on different gases hardly matters.

Similarly, pressure fields aren't limited to water, and their ability to crush submarines extends beyond water, too. It doesn't matter whether the sea is water, alcohol, or methane; if the pressure field is too strong, that's the end.

These examples illustrate that a field's behavior may be largely independent of the particular medium in which it arises. Bending fields are found in rubber and metal as well as in rock; mass density is a property of any material; cobalt can magnetize as easily as iron.

This independence of a field from its medium is extremely important for this book. On the one hand, it explains why scientists studying a field may sometimes choose to say little about its origin: its medium may be largely irrelevant. One can study general properties of wind without having to focus on a particular planet's atmosphere. On the other hand, it also explains why scientists may simply not know a field's origin; early experiments on the field may not reveal it. Dust storms were seen on Mars, and its wind inferred, long before the gases that make up its atmosphere were identified. Humans have long been aware that the Earth is magnetized, but the nature and properties of its core, where its magnetization is generated, are still not entirely clear. In the same vein, although scientists have a profound understanding of electric and magnetic fields, they still don't know whether these fields have a medium or what properties they might represent.

Conversely, *a particular medium may have many properties and therefore many fields*. A chunk of iron can have a mass density, a pattern of bending, a degree of magnetization, etc. Air (on any planet) has density, pressure, wind, and humidity; any ocean can have pressure and flow. Each of these properties can vary across its medium and change over time, as expected for a field.

If several of a medium's fields can be observed, we can gain clues about that medium's nature by studying how the fields interact—i.e., how they affect one another. For example, the interaction of a medium's pressure field, flow field, and mass-density field offer insight into whether the medium is a gas, a liquid, or a solid.[2]

As another example in which a medium's details might be revealed by its fields, imagine a large but paper-thin sheet of aluminum. It's so thin that it might not occur to you that it has a significant thickness, visible if you look under a microscope. Yet its three-dimensional nature is revealed by its fields.

Even as a two-dimensional sheet, it can be twisted and bent here and there, so it's clear that it has a bending field. However, if you carefully studied how the material behaves, you'd soon find that it has at least two other fields. Their origins are sketched in Fig. 33. One involves compression of the hidden three-dimensional structure of the aluminum. This compression field wouldn't be obvious if you imagined the aluminum as infinitely thin, but even a paper-thin sheet is a grid millions of atoms thick and can indeed be compressed. The other might be called a leaning field; it tells you the degree to which the stacking of the aluminum atoms, from the bottom to the top of the sheet, is tilted. By discovering the compression field and the leaning field, and perhaps others, and by studying how these fields behave and how they interact, you could begin to learn a great deal about their medium—for instance, that it's a three-dimensional solid.[3]

This example serves to illuminate a path that scientists have often followed in their research. While we might first discover a medium and then learn about its properties—its fields—sometimes it's the other way around. We may discover a field first without being able to comprehend or even identify its medium, leaving us ignorant of the field's ultimate origin. But as we study the field's behavior and discover more fields that interact with it, we gain additional information that allows us to dig deeper. Eventually we may come to understand its true nature and that of its medium.

Our universe has a host of fields, found everywhere, even throughout empty space. (I'll refer to them generally as *cosmic fields*.) We know very little about what they are or where they come from. The questions that I just

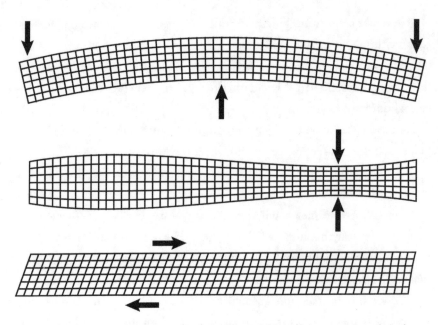

Figure 33: A thin aluminum sheet's bending field (top) is easy to see, but the compression and leaning fields (middle and bottom), much more obscure to a casual observer, reveal hidden details of the sheet.

outlined stare particle physicists in the face. It's tempting to imagine that these cosmic fields are properties of an everywhere-medium that makes up the essential structure and fabric of the universe.

But we must remain cautious. Such questions sound reasonable, but they might not be the right ones. It's possible that our extensive experience with ordinary fields could mislead us. That happened once in the past, and it took Einstein to set us straight, as we'll soon see.

Like particles, cosmic fields may or may not be elementary (as in "elemental"). Just as there are composite particles, such as protons, that are made out of other particles, there are composite fields that are made from other fields. Both particles and fields can be up-to-now-apparently-elementary; an elementary field is simply a cosmic field that has not, so far, shown any indication that it is composite.

Almost all the cosmic fields that we will deal with in this book are (apparently) elementary. Though composite cosmic fields have occasionally turned up in particle physics,[4] I mention them in this book for one reason only: the Higgs field might be composite. So far it seems elementary, and in most of this book, I'll assume for simplicity that it is. But we still know rather little about this field, and it is possible that experiments in the near term will reveal signs of compositeness. I'll return to this issue in later chapters when we explore current puzzles facing particle physicists.

In total, roughly two dozen up-to-now-apparently-elementary fields have been discovered by physicists, including the gravitational, electric, and magnetic fields as well as the Higgs field. The relation between elementary fields and elementary particles will be central to our story going forward.

The word *field*, like so many others, has meanings in English that differ from the one used by physicists. But the word is also popular in another dialect: that of pseudophysics, used to make ideas sound scientific even when they're not. Famous from science fiction is a "force field," used as a protective shield. Such a concept does not exist in physics despite the relation already alluded to between certain fields and certain forces. If you aimlessly wander the internet, you will find websites describing "quantum energy fields" and "consciousness fields." These aren't physics fields, either.

I once had a conversation about this with a woman who insisted that such pseudofields exist and that she could feel them.

"I know you physicists don't believe in these things," she said. "But perhaps you're too skeptical? After all, you can't see or feel magnetic fields, either, but you believe in them, I imagine."

I paused before replying. She was, from my perspective, missing the point. There's no dispute that real things exist that our senses can't easily detect. We can't see radio waves, for instance. Nevertheless, we know they're real; if they were not, we couldn't use them for long-distance communication. If X-rays weren't real, they couldn't be detected by photographic plates and used to help doctors set broken bones and treat cancer. Things that can be reliably and predictably useful are, almost by definition, real, and it doesn't matter whether our senses pick them up. Furthermore, what seems undetectable need not remain so, thanks to technology; machines and other simple devices often transform something we cannot sense into something we can.

And often, if fields are strong enough, no technology is actually required, as I pointed out.

"Magnetic fields, and other elementary fields that physicists talk about, are invisible, but they're not obscure and beyond human senses. No special expertise or belief system is needed for you to experience them. Your hair, for example, is highly responsive to a whole host of invisible fields."

Hair is great for detecting the wind field, obviously, but it can do much more. When there's a strong electric field around, as when you remove a wool hat in winter or are about to be struck by lightning, your hair will tell you: it will stand on end. As for the gravitational field, your hair detects it by hanging down rather than floating around aimlessly. We could even use your hair to detect the magnetic field; if we sprinkled your hair with sticky iron filings, it could respond like any compass.

Your hair can detect the Higgs field, too. If that field, currently switched on across the universe, were suddenly switched off, then your hair would explode—along with the rest of you. I think that would count as sensory detection, even though it would be extremely brief.

"Such methods will work on anyone's hair," I said to her. "They'll even work for a child with no scientific knowledge or aspirations. So even though many fields of physics are invisible, they *can* be felt reliably, with the right devices. They're not in fact beyond human senses, either mine or yours."

The important differences between physics fields and pseudofields are re-liability and clarity. Any physics field, or, at a minimum, its waves, can be observed and measured in a clear, well-defined, quantitative fashion, one that is straightforward to explain and, for anyone with sufficient resources, to repeat. If a field from pseudophysics were to pass these kinds of tests, it would be accepted as a physics field.

She wasn't impressed by this requirement. It struck her as far too fussy, the sort of thing that only closed-minded people would insist upon.

"You don't have to like or approve of scientists' standards," I responded, "but these are the ones we keep. In physics, we accept that we might oc-casionally fail to recognize something as real in return for confidence that *we never mistakenly accept an illusion.* That's how we ensure that scientific knowledge, limited as it might be, remains trustworthy. Everything we physicists claim to know has passed through a long chain of rigorous checks."

She thought for a long moment. Then she made an interesting observation. "For me," she remarked, "it's the exact opposite. I'd prefer not to reject anything that might be true, even if it comes at the cost of occasionally believing something that's an illusion. I mean, suppose I'm wrong, and consciousness fields don't really exist after all. It's pretty harmless; I haven't hurt anybody.

"I never really thought about the fact that for scientists, the attitude is the reverse. It explains a lot."

It does. And so we agreed to disagree.

14

Elementary Fields

A First, Unsettling Look

In this chapter, we make a first attempt to grasp how the cosmos works, trying to understand its fields using analogies with the ordinary fields of the last chapter. As we proceed, we will be retracing steps taken by generations of physicists over the past century and a half. Right from the start, our approach will encounter problems, and the further we go, the worse the difficulties will become. I tell you this now so that you won't be expecting a moment of sudden enlightenment. This chapter will end ambiguously, and we will escape to firmer ground only thereafter.

But we have to travel this route. To fathom our astonishing universe requires a deep appreciation of why the last century's greatest scientific minds have failed to make sense of it, and why it could justifiably be called "an impossible sea."

The basic message of the last chapter is that an ordinary field represents a property of an ordinary medium. In the chapter before that, we encountered the idea of a medium in the context of waves: air as the medium for sound waves, rock as the medium for seismic waves, and so on. Now we want to put this triad of concepts—medium, field, and wave—together.

We've largely established how the members of our first triad, air, wind, and sound, are related to one another. Wind is the flow of air. Sound waves are ripples in air. Air is the medium for sound waves in particular and for the wind more broadly. That's not quite all, though; the relation between wind and sound remains to be outlined. More on this shortly.

Similar relationships exist within the triad of rock, the bending field, and seismic waves. In a moment, we'll see another: iron, its magnetization field, and something called *spin waves*. A goal of this chapter is to explore how these triads might function in general, even for the fields of the universe.

As we saw in the last chapter, the underlying interpretation of an ordinary field may prove elusive, as was the case for the magnetization field of iron. When it comes to the elementary fields of the cosmos, our ignorance is that much greater. With one exception, none of the fields has an identified medium, and the properties represented by the fields are unknown. Worse, it's not even clear whether the elementary fields have media at all, despite what we would expect from our experience with ordinary fields. Perhaps elementary fields break the ordinary rules; why not?

When a field is known but its corresponding medium is not, we may need to adopt an unfamiliar perspective in order to talk about it. As an illustration, let's again consider the triad of air, wind, and sound. Suppose you'd never heard of air or guessed its existence. You'd know about wind, having felt it and seen its impact on the world, but you'd be mystified as to its cause and meaning.

Perhaps this isn't so hard to imagine. As a friend of mine remarked, "That's what it's like for little kids, isn't it? I still remember a horrific storm from early childhood that snapped many of our neighborhood's trees. I didn't understand at the time that wind is simply the movement of air. Instead, I had some weird idea about it being an invisible animal."

"That's a good point," I said. "And in that sense, when it comes to electric and magnetic fields, scientists are still toddlers."

She chuckled. "Yeah, well, I've always thought of you physicists as grown-up children."

"Oh, totally," I laughed. "Infinite curiosity, lots of naivete, and very little knowledge in the grand scheme of things."

Even without a conception of air, scientists could discover a great deal about the wind through careful observations using an array of wind meters. They could infer that a storm's wind twists counterclockwise in the northern hemisphere and clockwise in the southern. They could also learn that the wind often points upward in thunderstorms, explaining why their clouds grow so high. From these studies, they could guess formulas that relate the wind field to the atmosphere's pressure field, humidity field, and temperature—all of which can be measured without knowing anything about air. These formulas would eventually help the scientists recognize something more fundamental: that the wind field represents the flow of a gaseous medium.

But before they came to this understanding, what would these air-ignorant scientists think of sound? One might doubt that they'd be able to understand it, since sound involves waves in air. But this would not stop them. They would conclude, from their experiments, that *sound is a wave in the wind.*

"Are you saying they'd be making a mistake?" my friend asked.

"No," I replied, "they wouldn't be wrong. Sound is *both* a wave in the air *and* a wave in the wind. Those are two perspectives on the same thing. They'd just be taking the perspective we find less familiar, and doing so out of necessity."

One wave, two perspectives. In a moment, I'll illustrate how this works for air, wind, and sound, and in a few pages, I'll give you another example, partly to help the idea settle in and to emphasize how general this situation can be.

When we describe sound as we usually do, as a wave in the medium called air, we are adopting a *medium-centric* perspective. In Fig. 34, I've depicted a snapshot of a sound wave as it moves to the right. The medium-centric perspective focuses on what the air is doing: it is more concentrated in some places (let's call those the crests, drawn as darker regions with vertical arrows pointing at them) and less so in others (the troughs, drawn as lighter regions).

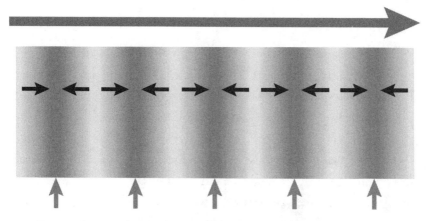

Figure 34: A simple sound wave traveling to the right (long gray arrow). It has crests (more concentrated air, in darker shading and indicated with vertical arrows) and troughs (less concentrated air, in lighter shading). The wind field is rippling, too (black arrows).

Meanwhile, the wind is also rippling. For air to become more concentrated somewhere, it must flow into that location. Similarly, the air must flow out of places where it is becoming less concentrated. In short, wind blows into crests and out of troughs. These flows are indicated by the black horizontal arrows in the figure. At the center of a trough or crest, the wind field drops to zero.

So in this snapshot, the wind has its own ripples—rightward, then zero, then leftward, then zero, then rightward again—over and over across the sound wave. This whole rippling pattern of the wind field moves to the right in lockstep with the rippling pattern in the air. So if we take a *field-centric* perspective, focusing on the wind instead of the air, we could describe this sound wave as ultimately a wave in the wind field—a wave in a property of the medium called air.

There's only one wave here, not two. A ripple in the air can't exist without a ripple in the wind, and vice versa; that's what Fig. 34 illustrates. So our two perspectives are simply two ways of looking at a single wave: one emphasizes the air, the other emphasizes the wind. Which one we choose is up to us. For those of us who know about air, wind, and sound—medium, field, and wave—both perspectives are equally good, even though the latter may seem a little odd and unusual at first.

But now imagine that you have only part of this picture. You're a scientist who knows nothing about the existence of air or of the rippling pattern illustrated by the shaded regions in Fig. 34. All you can measure are the black horizontal arrows; all you can know is that the wind is rippling and that its ripple is traveling to the right.

This is the challenge for those poor scientists: they can measure the wind, but they don't know more fundamentally what it is that they are measuring. Nevertheless, using experiments and observations, they can learn what the wind field *does*: it knocks trees down when it's strong; it can spiral around in tornadoes; it can push on walls and eardrums. They can write detailed formulas for its behavior. They can learn a great deal about sound waves, too: amplitude, frequency, speed, and effects on human ears. They can observe that when you and I hear sound, there's always a wave in the wind field that's responsible. And so they will naturally adopt the field-centric perspective and describe sound as a traveling wave in the wind.

In a sense, this perspective is incomplete. Conceptualizing sound waves purely in terms of wind leaves out their fundamental relationship with air. But until the scientists know what's missing from their knowledge of sound and wind, it will have to do. At least it's clear and specific.

By contrast, imagine them trying to use the medium-centric perspective. Based only on the black arrows, what could they say? What could they even think? *Sound is a wave of some sort in an unknown medium, and somehow that wave also makes the wind field ripple.* Okay, that's true, but it's too nebulous to be useful. Maybe they'd give that medium a name: the soniferous aether. *Sound is a wave in the soniferous aether.* Would that really help them? All it would do is repackage their ignorance into fancier-sounding language; they still wouldn't know what this aether is and how wind relates to it. They could try to guess what the soniferous aether might be, but without supporting evidence from experiments, such speculation would be of limited value. So in this situation, the medium-centric perspective would be both vague and speculative, and it is doubtful they would use it.

To illustrate how common this situation is, I'll give you another example. Let's take an iron magnet, illustrated at the top of Fig. 35. Its little atomic magnets all point in the same direction, assuring that the magnetization field, which characterizes the degree and direction of atomic alignment (see Fig. 32 in the previous chapter), is constant across the iron. If disturbed, this magnet may exhibit what is called a *spin wave*, as depicted at the bottom of Fig. 35. Within a spin wave, in contrast to a sound wave, the atoms do not change position as the wave passes. Instead, as this wave travels along, it is the atoms' *orientations* that rock back and forth in a rippling pattern.[1]

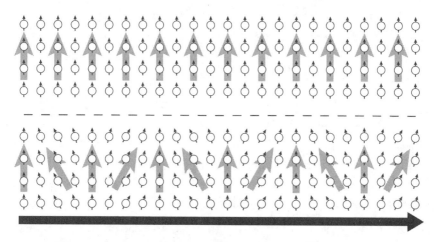

Figure 35: (Top) Magnetized iron; the atoms are aligned, as is the magnetization field (gray arrows). (Bottom) As a spin wave moves to the right through magnetized iron, the atoms' orientations and the magnetization field rock back and forth.

One interesting aspect of this wave is that unlike the water in an ocean wave or the rock in a seismic wave, the iron doesn't change shape in a spin wave. Only some of its internal properties change. So this isn't a wave you could see. It's hidden, microscopically, inside the iron. That's an important point: most of the waves we observe in daily life change their medium's outward appearance, but this need not be the case, especially where the universe is concerned.

I have just described a spin wave from a medium-centric perspective, using my knowledge that the magnet is made of iron, the iron is made of atoms, and the atoms' orientations are waving. But if I didn't know atoms have orientations or perhaps even that atoms exist, I'd be hard-pressed to describe the spin wave this way. It would be even worse if the iron were completely invisible to me and I couldn't even know whether it was a solid, a liquid, a gas, or some other exotic material.

However, instead of focusing on three things I don't know about—the iron, the atoms, and their orientations—I could focus on something I can actually observe: the magnetization field. Since the atoms all rock back and forth together in a spin wave, the magnetization field rocks, too, as illustrated by the swaying gray arrows in Fig. 35. As was also true for the sound wave, there's only one wave here: the magnetization field can't wave unless the atoms' orientations do, and vice versa.

As the spin wave passes by, needles placed on the iron's surface will respond to the waving magnetization field by rocking back and forth themselves. Observing the needles' behavior, I could detect, infer, measure, and study the spin wave as a wave in the magnetization field, all without any knowledge of iron and its atoms.

In this way, I could characterize the spin waves from a field-centric perspective. My ignorance of the relationship between the magnetization field and the detailed structure of the medium that underlies it would make it difficult for me to employ the medium-centric perspective. Too bad for me. But I could still do good science with what I know.

When dealing with invisible, remote, or otherwise obscure media, this situation is not unusual. In the best case, we understand the full triad: medium, field, and wave, and then either the medium-centric or field-centric perspective will do. But if we know little or nothing about the medium, we often turn out of necessity to the field-centric perspective.

When it comes to light waves, and almost all other waves of the universe, the field-centric perspective is the only one we currently have. Our triads are partial. Not only have we never observed the corresponding medium (or media), but we're not even assured of its (or their) existence. That's why physicists, and we in this book, have no choice but to adopt the field-centric view, incomplete as it may seem…and incomplete as it may indeed be.

With this in mind, let's now turn our attention to some of the best-known elementary fields. Historically, the three elementary fields that we've encountered so far were the first that scientists recognized as such, through the forces they create: the electric field, the magnetic field, and the gravitational field. But in fact, as understood since Einstein's day, the electric and magnetic fields are really two aspects of one field, known as the *electromagnetic field*.

In the nineteenth century, profound connections between electric and magnetic phenomena were gradually uncovered. For instance, an electric field can create an electric current in a wire, but that current then causes the magnetic field to loop around the wire. Or consider a moving magnet. A magnet activates the magnetic field around it. But if you move rapidly past that magnet, you'll see the field around it as part magnetic, part electric.

In 1831, Michael Faraday, who invented the idea of a field as used in modern physics, discovered that a changing magnetic field leads to a response in the electric field. Thirty years later, James Clerk Maxwell realized the reverse is also true: that a changing electric field makes the magnetic field respond, which causes the electric field to react in turn. This creates a chain reaction, in which a ripple in either field generates a ripple in the other. The result is a wave that is both electric and magnetic, or, simply, *electromagnetic*.

Moreover, Maxwell showed that the velocity of this combined wave matched the most recent measurements of light's travel speed. He then proposed that light, both visible and otherwise, is an electromagnetic wave—a wave in the electromagnetic field.

Just as air serves as the medium not only for sound waves but for wind more generally, the luminiferous aether, if it exists, would act as the medium not only for light but also for the electromagnetic field as a whole. Maxwell, like all nineteenth-century scientists, assumed that the aether exists, and he made some proposals as to what it might be like and how the electromagnetic field would stem from it. But experiments soon disproved his suggestions.

The corresponding questions for the gravitational field's medium were long unanswered, too; Newton's formulas for gravity offered no clues. But when Einstein began revising physicists' understanding of relativity in 1905, he knew right away that Newton's formulas couldn't coexist with his novel ideas. It took him ten years of hard work and wrong turns before he found new formulas, called *general relativity*, that could both describe gravity and be consistent with his update to Galileo's principle. According to these formulas, empty space should be understood as a medium, and the gravitational field, much more complicated than for Newtonian gravity, tells us how this medium is warped.[2]

Einstein thereby gave us our first concrete example of how an elementary field could be described as a property of a medium. It's still the only one. But it already teaches us an enormous amount about the universe, as we'll see.

Air and rock and water and many other ordinary media can have waves. Can empty space ripple, too? Einstein himself became confused about this technically challenging question. But his formulas do indeed predict the existence of waves of empty space, the "gravitational waves" that I've mentioned a couple of times.[3]

Gravitational waves aren't an abstraction. If a large one passed, we'd know it. Though we don't have sensory organs dedicated to these waves the way our ears and eyes target sound and light, we could still feel them if they passed by. They would stretch and squeeze our bodies, as you might guess for waves of space. Near the collision of two black holes, each with the rest mass of a few Suns, we'd easily feel our shapes being distorted. In fact, we could be ripped apart if we were too close!

If that's so, why haven't you or I ever noticed one of these waves? Even the largest volcanic and nuclear explosions on Earth are far too small; the gravitational waves they make aren't detectable by even the best scientific equipment. To make a powerful gravitational wave requires an even greater upheaval, such as the colliding and merging of two large black holes. Such violent events are rare and unlikely to occur within any nearby galaxy during any human lifetime. But there are many more galaxies far away than close by (for the same reason that most humans live far from you—there's more living space outside your neighborhood than inside it), so when black holes do merge, it is most often in a remote galaxy. The gravitational waves produced in the merger have large amplitudes initially, but they spread out as they travel, and their amplitudes shrink like those of ripples in a pond. By the time they reach Earth, their amplitudes are simply too small for humans to perceive them.

Even so, scientists have confirmed their existence. About fifty years ago, astronomers Joseph Taylor and Russell Hulse discovered two neutron stars closely orbiting one another. Measuring the duration of each orbit carefully, they found a slow trend toward briefer orbits. The trend perfectly matched the predictions of Einstein's gravity formulas, confirming that gravitational waves were being emitted by the pair. For this achievement, Hulse and Taylor were awarded the 1993 Nobel Prize in Physics.

Then in 2015 came direct observation of gravitational waves. An experiment called LIGO (the Laser Interferometer Gravitational-wave Observatory), packed with the latest technical wizardry, managed to detect the squeezing and stretching of the Earth caused by the gravitational waves from two black holes colliding and merging into one.[4] You and I felt nothing because the wave's amplitude was tiny; as it passed, Earth's diameter changed by less than an atom's width. It might seem incredible that anyone could ever hope to observe an effect as diminutive as that, but my colleagues who built and run LIGO are wizards, complete with degrees from quantum wizard school. Nobel Prize? You bet—2017. And what was extraordinary is now commonplace; aided by major technological upgrades, LIGO and its partners, Virgo and KAGRA (Kamioka Gravitational Wave Detector), observe gravitational waves from extraordinary events in the universe several times a week, as of 2023.

So the triad of medium, field, and wave has appeared again. Analogous to air, wind, and sound waves as well as iron, its magnetization field, and its spin waves, we now have empty space, the gravitational field, and gravitational waves. It seems, from this viewpoint, that much as the Earth is made of rock and its oceans are made of water, the universe, too, is made of a medium: empty space itself.

Yet something is amiss. There's something uncanny about empty space, apparent even in the small gap that separates your eyes from the page or screen on which you are reading this sentence.

Of course, what separates you from the page or screen is filled with air, but that's a distraction. First, the atoms that make up the air are mostly empty space. Second, if we gave you a space suit and pumped all of the air out of your room, you and your book would now be separated by truly empty space, but since light would easily cross the gap, you could still read the book just fine.

I mentioned this to a friend, who asked, "Two feet of emptiness? What's so strange about that?"

"Well, it's very different from the nothingness that most of us would imagine emptiness to be," I pointed out. "Empty space is as empty as possible, sure. But if empty space were really nothingness, how could it become warped and create gravity? How could there be waves in it? How could the universe expand? These properties make it seem more like rubber or fabric than nothingness."

"Yeah, I agree that's weird," he conceded. "But what's your point?"

I leaned forward. "We don't have any trouble seeing and feeling rubber or fabric," I replied. "Why is it that we've never been able to detect empty space?"

Based on similar questions that we've already come across in this book, you may naturally guess that we face yet again a failing of our senses. The mere fact that we can't see or feel the substance or essence of space doesn't prove anything; we humans are limited. Perhaps this is just one more thing we can't detect directly.

Yet the problem is much more profound than that. It's not just about human senses; no scientific instrument of any type has ever detected a materiality—a somethingness—to empty space, either. In fact, if one ever succeeded, it would be revolutionary...because it would violate a deeply cherished principle.

It would violate the principle of relativity.

Whoa.

That's right. Although empty space is apparently a medium that serves as the universe's fabric and brings about gravity's pull and gravitational waves, the principle of relativity requires that it be undetectable and that it forever remain so.

Well, now. This is a lot for a human brain to digest. We need to step back, take a deep breath, and try to figure out what our crazy universe is trying to tell us. That's our task for the rest of this chapter.

Afterward, I'll take you onward to fields and particles and to what we're all made of. But what I've just revealed should be sobering. There's still so much, even about the very basics of being, that lies beyond our grasp.

14.1 A Medium for the Universe

Empty space is a medium unlike any other. It's completely transparent—light can cross it for billions of years without dimming or slowing. It's completely permeable—you and I and our planet go right through it, at 150 miles per second relative to our galaxy's center, without any consequences. Location within it is unmeasurable and steady motion through it is undetectable—that's the principle of relativity. And it's everywhere...or maybe we should say that "everywhere is *it.*" Scientists often talk about it as though it's the container for everything that exists in the universe.

Compare this with ordinary media that we've encountered. Clearly rock and iron are neither transparent nor permeable. Water's transparent to a degree, but it's clearly not permeable, as anyone who's done a poorly executed dive or a purposeful belly flop into a pool will tell you. Air is transparent, at least to visible light, and we move through it easily. But objects that slam into the Earth's atmosphere at high speed find it far from permeable; that's why spacecraft returning to Earth need heat shields and why most meteors vaporize long before hitting the ground.

Ordinary media can be escaped, too. We can choose whether to be inside or outside the Earth's rock, the ocean's water, and the atmosphere's air. When we're in them, we push their material out of the way; for instance, when you swim, the water flows around your body, not through it. This is true of any familiar medium made of ordinary material. Empty space is different. It's an *everywhere-medium*, a medium of the cosmos. We're always inside it, and it's always inside us. There's no place we can go where we can look at empty space from a place that doesn't have any. That would mean going outside the universe.

A further crucial difference is that empty space cannot be collected, carried around, and stored for detailed investigation. Compare this to seawater; we can put some in a bottle and examine it in our kitchens. An intelligent fish could fill a bottle with air and study it within the ocean. Either of us could bring large rocks into our water-filled or air-filled laboratories. For that matter, we could take bottles of air, water, or rock into empty space and inspect them there. The reverse is not true. If you empty a bottle of all its contents, the bottle is in a sense full of empty space, but the empty space has not been bottled the way water can be. What's inside the bottle is empty space, but what's outside the bottle is mostly empty space, too. Even the bottle walls are mostly empty space. You cannot fill a bottle with empty space and take it somewhere; you haven't captured any. Nothing stops it from gliding right through the walls, from inside to outside and from outside to inside, as you carry the bottle to your air-filled (but mostly empty space) laboratory.

When you drive with your windows up, your car is like a bottle. The car's atoms, impenetrable to other atoms, block the outside air from entering the car and block the air inside from exiting. You carry your own air with you; that's why you feel no wind inside the car even at highway speeds. But the car isn't impermeable to empty space. As it moves, it doesn't push the empty space in front of it out of the way, nor does it carry the empty space inside the car along with it. The car's atoms just move smoothly through and across empty space.

The same is true when you walk down a street; the air gets out of your way, but the space doesn't have to. It just goes right through you, and you just go right through it.

These curious and unsettling features of empty space must also be true of the luminiferous aether. If it actually exists, it, too, must be an everywhere-medium; not only are light waves pervasive outside material objects, but the electric field is the intermediary holding every atom together. Like empty space, this aether is transparent and permeable, generates no resistance as you move through it, and can't be bottled or pushed out of the way. You can coast smoothly through it, and it through you, without effort.

In fact, these same characteristics must hold for any elementary field's medium, and more generally for that of any cosmic field, elementary or not. Since any cosmic field exists everywhere in the universe, its medium, if it has one, must be an everywhere-medium. That medium must be transparent and permeable. Why? Because *even a single nontransparent or nonpermeable everywhere-medium would render the entire universe opaque, impermeable, or both.*

I've suggested that detecting anything material-like or substance-like about empty space or the luminiferous aether would violate the principle of relativity. Now I need to explain why this is so. Toward that end, let's consider some established methods that work well for detecting an ordinary medium. We'll soon see why, despite all odds, they fail us.

In the heart of Seattle, where I taught for several years, lies Lake Union. The lake is crammed with sailboats and motorboats on summer weekends. On its east bank rises Capitol Hill, with a highway halfway up its steep slopes bringing cars to and across a bridge that joins the northern and southern halves of the city. In mirror image, Queen Anne Hill dominates the western shore, bearing another north-south highway with a bridge of its own. At the water's southern edge stretch industrial areas and the skyscrapers of the city's urban core, while a few thousand feet overhead, jets approach or depart from Seattle's airport, which lies 10 miles south. In the middle of all this, seaplanes take off and land directly on the lake, avoiding the boats using strategies that I could never quite figure out. It's a graceful ballet of modern transportation.

If you've ever been on a similarly crowded lake or harbor, with watercraft gliding every which way, you know how you can lose your sense of reference, making it almost impossible to figure out which way you're going and how fast. To some extent, this is another instance of the principle of relativity;

you can't tell your motion by feel, and trying to do so by eye is confused by all the other boats.

However, it's easy to figure out whether you're stationary *relative to the lake* and, if not, how you're moving. The simplest trick? Just stick your hand into the water!

If you feel the water pull on your hand, and your hand leaves a wake, then you know you're in motion. For instance, if your hand is pulled (and the wake points) toward the south, then you're traveling north relative to the lake. The faster you go, the more drag you feel and the stronger the wake.

Similarly, if you put your hand out of the window of a moving car, it will be pulled backward, more powerfully the faster you're traveling. As you are skateboarding or biking, touch the ground lightly with a stick; the stick will be dragged backward along the ground (from your perspective) and will leave a trail behind it.

Let's call this general approach the *drag method* for determining your motion with respect to an ordinary medium; see Fig. 36. You simply push something into or against the medium, and your speed and direction relative to the medium are revealed by the drag you feel or the wake you see. Moreover, if you stop running your engine, kicking your feet, or using your sail, you will find yourself slowing down until you've become stationary relative to the medium, at which point the drag and wake will cease.

What happens when you use this strategy to try to detect empty space or the luminiferous aether? Nothing. These everywhere-media cannot create any drag. If they did, they would do so even inside an isolated bubble, where you could use the drag's presence, strength, and direction to determine how you are moving through empty space. That would violate the relativity principle.

Thus, if Galileo's principle is correct, drag from any everywhere-medium is illegal, and so there's no hope of directly verifying the existence of such a medium this way. We will need a different approach.

Fortunately, there are other methods for detecting the presence of a medium. The best of these, well suited to our current purposes, is what I'll call the *wave speed method*. Since both empty space and the luminiferous aether

(if it exists) are well known for their ripples, and since waves are characteristic of all the elementary fields of the universe, this is a method well worth
trying.

The wave speed method can be used on any medium that's uniform and
behaves the same way in all directions. That seems to be true of empty space
and all other everywhere-media, whose fields show no preference for any one
direction or location.

Here's how it works, again illustrated in Fig. 36. First, measure the speed
of the medium's waves as they come toward you from one direction. Then
compare the answer with the speed of identical waves coming from any
other direction. If you're stationary relative to the medium, the speeds of the
waves from all directions will be equal. If you're moving, they won't be.

Figure 36: How to determine motion relative to an ordinary medium, as seen
by someone stationary (left) or right-moving (right). (Top) The drag method,
showing (right) the wake created by an oar moving through the water. (Bottom)
The wave speed method, with wave speeds indicated by black arrows.

For example, suppose you're in a boat and want to know if you're moving
through the water. Have someone drop a rock just north of you, and observe
the speed with which the ripples from the splash move southward toward
you. Then have someone drop a similar rock just south of you, and make
a similar observation of the northward-moving ripples. Do the same with

the ripples from rocks dropped to your east and west. If you are stationary relative to the water, the waves from all the splashes will approach you at the same speed.[5]

But suppose instead you're moving north through the water. Then northward-moving wave crests coming from south of you—from behind you—will have trouble catching up. From your perspective, their speed will be reduced compared to the speed of southward-moving wave crests. That's the clue that reveals that you're moving through the water; the difference between the waves' speeds of approach tells you your motion's speed and direction.

A student in one of my classes told a relevant story. "We were on a whale-watching trip and were moving slowly toward a distant beach. Waves heading for the beach were passing us. Then the captain turned the engine on high. Soon we were moving at the same speed as the waves, which appeared as though they were frozen in place!"

As you move faster through the water, a wave traveling in the same direction will seem to move more slowly. If you perfectly match your boat's motion with that of a wave, you'll see its crests and troughs as completely motionless. If you go even faster, you will pass them, and the wave will appear to move backward.

You can similarly catch up with and overtake sound waves in a supersonic aircraft. You'll always be able, with a powerful enough engine, to outpace any ordinary wave in any ordinary medium. But no matter how hard you try, you can never catch up with light waves that have passed you. The faster you try to go, the more they will seem to recede from you, always at exactly the same speed: c, the cosmic speed limit.[6]

At this news, one of my adult students laughed nervously. "Umm...I've had dreams where someone steals my briefcase or my phone and I run after them, but the faster I run, the faster they run, and they keep getting further and further away. And light is like that?"

"Kind of," I replied. "And if you try to run away from light that's approaching you, it's hopeless; you'll never escape it."[7]

"So you're telling us that reality is something out of a nightmare," he lamented.

I grinned ruefully and nodded. "I suppose I am, yes."

Like the drag method, the wave speed method fails to reveal any motion through the luminiferous aether. Consequently, it cannot be used to confirm whether the aether exists. No matter how you move, light waves approach you from all directions at the same speed, as though you're always stationary relative to light's medium.[8]

This is the way it has to be, as we can see by considering the alternative. Imagine that the wave speed method worked in our universe. Then, when moving northward through the luminiferous aether, I'd see light waves coming from the north move past me faster than those from the south. The difference would be evident even in an isolated bubble, using two flashlights at opposite ends of the room. From the discrepancy between the speeds of the light from the two flashlights, I could deduce my motion across the aether, violating Galileo's principle. That's why, in order to preserve the principle of relativity, the speeds of the light from the two flashlights must always be equal, as Einstein proposed.[9]

The same would be true if you tried the wave speed method using gravitational waves. All such waves from all directions always approach you at the cosmic speed limit, no matter how fast you move in any direction you choose. You can never catch up to them. You can never escape them. (I should hasten to say that no one has done this experiment directly. But what we've seen of gravitational waves has confirmed that they adhere in great detail to Einstein's formulas for them, which require that they have the same nightmare property as light waves.)

Empty space and the luminiferous aether refuse to give us any hint of our motion through them, or indeed any direct hint that they exist at all. They always act as though we're stationary. We may say they are *amotional*—no matter how we move, our motion relative to these media has no measurable consequence and seems to have no meaning at all.

Looking back, we can see we could have guessed this would happen. If these everywhere-media weren't amotional, then we could measure our speed relative to them. But because they're everywhere, they're found even inside isolated bubbles. And so, even inside those bubbles, there would be measurements we could make that would reveal our motion. This can't be so if Galileo's principle is true.

There's nothing amotional about ordinary objects or ourselves. We do move. But in specifying our speed and direction, we must do so relative to another object. Since any object will do as well as any other, it's our fate, and that of all objects in our amotional cosmos, to be poly/omni/ambimotional. There's no best, truest way to define one's speed.

The failure of the wave speed method to show any sign of the universe's everywhere-media is even more traumatic for our understanding than was the absence of drag. You could perhaps imagine a magical substance that exerted just the tiniest drag, so ethereal that it could let macroscopic matter pass through it almost completely unscathed.[10] But the problem of wave speeds can't be imagined away. It is literally impossible for ripples in any ordinary medium, made from any ordinary material or anything like ordinary material, to be amotional or to have the nightmare property. It's logically inconsistent. Just ask my students.

"If every driver and sailor and pilot around Lake Union used the wave speed method," I claimed to my class, "each would find that light waves from all directions approach at the cosmic speed limit. None of them would detect any motion relative to the luminiferous aether, despite the fact that they all are moving relative to each other."

One of the students gave me an incredulous look. "So even though the passengers in an airplane flying overhead are obviously moving with respect to us, they're just as stationary with respect to light's medium as we are? That makes no logical sense!"

I smiled grimly. "For an ordinary medium, you're right: it makes no sense. If I'm stationary in the water and you swim past me, it's impossible for you to be stationary in the water also. But when it comes to a medium that fills the whole universe, the logic you're using has a subtle loophole. The issue involves how we measure and interpret time. As guessed first by Lorentz and put on a firm footing by Einstein, the passage of time is perspective-dependent. How often a clock seems to tick depends on how fast the clock is moving relative to the person observing it. There's a similar perspective-dependent distortion of distances.[11]

"This is critical, because you determine a wave's speed by measuring the amount of distance covered by the wave over a certain amount of time. Once time and distance get messed with, the way you measure the speed gets messed with, too.

"Your perfectly reasonable logic implicitly assumes that every observer measures speed using a clock that we all can agree on. This, Einstein realized, is not the case. Each of us, when we use our own clocks to apply the wave speed method and check our motion relative to empty space or the luminiferous aether, comes to the conclusion that we're stationary. There's no experiment that anyone could ever do that could demonstrate that some of us are right and the others are wrong. The wave speed method never reveals any form of motion or any sign of a medium for either the gravitational or electromagnetic fields.

"This could never happen in an ordinary medium such as water or air. An ordinary medium has no hope of affecting time and space! Only the universe itself can do it."[12]

"But why does any of this matter?" asked a third student. "We can just *see* the medium for the gravitational field."

"Can we actually see empty space?" I cautioned him. "We can *infer* the empty space between objects as well as the empty space inside them—the space they occupy. But we don't see the space itself. We see through it. We see the objects and infer that the space must be there. In fact, it's worse than that: our eyes merely detect light from those objects. Only later do our brains construct a picture of the world for us, in which there are objects and empty space between them."

"Can't we feel it?" another asked, waving her hands about. "I can feel my hand moving around in it."

"You feel your muscles flexing, and you probably can feel the air moving around on your hands. But again, we're moving through space at 20 miles per second around the Sun and at 150 miles per second relative to the center of the Milky Way. And you don't feel that, do you?"

There were no further protests.

No less weird, though perhaps less nightmare-inducing, is what happens if you try to retreat from light waves that are already receding from you. They'll seem to slow down.

You'd naturally expect that if light is moving away from you to the south at the speed c, and you start moving north, then the distance between you and the light will grow at a speed faster than c. But that's not the case—at least, not from your perspective. As you see it, the light will slow down in just such a way that the distance between you and the light still grows exactly at the cosmic speed limit.[13]

No matter how fast you chase or flee light waves, they will always seem to approach you or recede from you at the speed c. Bizarre as this may seem, Einstein's point was that it's required by Galileo's principle. If light waves did anything else, you could use them to determine your own rate and direction of motion. If you're moving steadily, light must behave just as though you're not moving at all, and so the speed with which it approaches or recedes from you from any direction must always be the same.

This is utterly different from sound. We describe "the speed of sound waves" as about one-fifth of a mile per second at sea-level pressure. But if all speed is relative, then what does this mean? Implicitly, we mean that sound's motion is to be measured relative to its medium; sound waves move at one-fifth of a mile per second *relative to the air*. But from the perspective of someone at our galaxy's center, viewing the sound waves traveling from you to me as carried along with the Earth and its atmosphere, those same sound waves would instead be moving along with us at a speed of roughly 150 miles per second. Only observers who see the atmosphere around them as stationary will view any and all nearby sound waves as moving at the standard speed of sound.

But by "the speed of light waves," we must mean something else. We can't mean light's speed relative to the luminiferous aether, because speed relative to an amotional medium has no meaning and can't be measured. Indeed, we mean the waves' speed as measured by any and all observers. Speed may be relative, but the speed of light waves, from everyone's perspective, is always the same. Though this seems logically impossible, and common sense may protest at this seemingly absurd claim, the relativity of times and relativity of distances save the day. They precisely compensate for each other whenever we use our own rulers and our own clocks to measure the motion of a light wave.[14]

This all sounds like something out of science fiction or a dream. But its reality is something our cell phones take for granted as they guide us through unfamiliar towns. GPS navigation requires extremely precise timing, and the differences between Einstein's formulas and Newton's older ones are too large to ignore. If you disputed Einstein, you'd soon end up in the wrong neighborhood or driving off a cliff.

Unlike gravitational waves or electromagnetic waves, the waves of many other elementary fields move below the speed limit. That's true for the Higgs field, for instance.

A student raised her hand. "Does the Higgs field's medium have a name?"

"Nobody's given it one," I replied.

"Higgsiferous aether?" suggested another student.

"Delightful," I grumbled dryly. But as often happens, the first proposal sticks, no matter how unfortunate.

"Since the Higgsiferous aether has waves that move below the speed limit," continued the first student, "does that mean you can catch up to them?"

"Yes, it does," I affirmed. "And pass them."

"So it's more like an ordinary medium?"

I shook my head. "It can't be. Remember, any medium for any cosmic field exists everywhere, even inside an isolated bubble. If any such medium weren't amotional, you could measure your speed relative to it, making your motion detectable. That would violate Galileo's principle."

Amotionality requires that the Higgsiferous aether can't exert any drag, no matter what the Higgs phib might claim. It must be as permeable as empty space and the luminiferous aether. Still, it differs from these media because its slower waves don't have the nightmare property.

Nevertheless, there's something miraculous about Higgs waves. Though they can approach or recede from you at many possible speeds, their range of motion is restricted. Higgs waves passing by from any direction cannot exceed the cosmic speed limit—and this is true from every observer's perspective.[15]

At first, that may sound unremarkable, analogous to the statement that ordinary passenger planes can't exceed the speed of sound. But the latter statement is not perspective-independent in the slightest; from the point of view of someone at our galaxy's center, all airplanes move at well over a hundred miles per second as they are carried along with our planet. By contrast, Higgs waves, and those of many other elementary fields, are limited to speeds below c as seen from everyone's perspective, no matter how the observer or the waves are moving. When you think about that carefully, it's a bit creepy. Speeds are relative, and yet these waves' motions are subject to intrinsic, perspective-independent constraints.

Yet again, this surprising behavior, logically impossible were it not for Einstein's novel conceptions of space and time, is required by the relativity principle. If the range of allowed speeds for Higgs waves changed depending on how fast you were moving, you could use a measurement of that range, even in an isolated bubble, to learn your own motion across the universe.

14.2 History, Aether, and a Missing Cat

Physicists came to this understanding (or lack of understanding) of the cosmos after a long struggle. In a sense, the struggle is not over.

The puzzles of the luminiferous aether troubled them for much of the nineteenth century. On the one hand, as the Earth spins and orbits the Sun, it carries us through the luminiferous aether without losing its atmosphere or slowing down. That suggests that the aether's interaction with ordinary matter is very weak. And yet the interaction of light waves with ordinary material isn't weak at all; that's why most objects are opaque. How does the medium, when inert, manage to leave ordinary material alone, even though its ripples do not? It appeared self-contradictory.

Still, it seemed clear that the wave speed method should reveal the existence of the aether, just as it would for any other ordinary medium. This difficult measurement became feasible in the 1880s, when Albert Michelson, a young leader in new experimental techniques, developed a device called an *interferometer*. An interferometer can measure differences in wave speeds with extreme precision. (There's one at the heart of modern gravitational wave detectors; hence the *I* in LIGO.) In 1887, with Edward Morley, Michelson built an interferometer powerful enough to detect the Earth's motion through the aether, even in the worst of circumstances. His approach differed in detail from the wave speed method I've described, but conceptually the issues were the same.

Yet no effect was seen. Attempts were made to attribute this surprising result to the properties of material objects as they move through the luminiferous aether. Perhaps, it was suggested, Michelson and Moreley's interferometer was distorted by its motion through the aether, its shape altered in such a way that the effect they'd expected to find was perfectly erased. Inevitably, any such notion would require abandoning the relativity principle. If the intrinsic structures of objects changed depending on how quickly they traveled through light's medium, then steady motion through this medium would be different from being stationary. One's motion through the aether would be detectable, even in an isolated bubble, by making precise measurements of nearby objects' shapes.

As a graduate student and patent clerk, Einstein was reading the papers by the famous physicists, including FitzGerald, Lorentz, and Poincaré, who were on the verge of setting aside Galileo's principle. He studied the math they were inventing in their struggles to explain why the wave speed method fails to reveal the aether. Then he cut the proverbial Gordian knot. He pointed out that Galileo's principle could be saved as long as the speeds of electromagnetic waves from all directions seem always the same, no matter how fast you're moving. This in turn requires that light satisfy the nightmare property and that its medium be undetectable in any ordinary way. To make the nightmare property logical, space and time must behave very differently from our commonsense expectations.

Repurposing equations that Lorentz had invented and giving them a very different interpretation than that of their originator, Einstein showed that these ideas make conceptual and mathematical sense. Rather than material objects changing shape as they move through the aether, which would violate Galileo's principle, what changes is the way we measure distances and durations, our basic methods for making sense of space and time.

Einstein's logic would then have required that the luminous aether be amotional. But instead, Einstein argued that it simply doesn't exist. With space and time and relativity having made it impossible to detect the aether using the drag method, the wave speed method, or any other similar method, there's no experimental evidence that it exists at all. That's why Einstein took the view that we ought to dispense with the very idea of it. After all, no self-respecting ordinary medium could possibly be amotional and nightmarish.

This is a breathtaking proposal: that light waves are utterly unlike sound waves, seismic waves, pressure waves, or any other waves we know of. Like the Cheshire Cat's grin, Einstein implied, light waves can exist without any medium at all.

> "Well! I've often seen a cat without a grin," thought Alice, "but
> a grin without a cat! It's the most curious thing I ever saw in my
> life!"[16]

Yes, a grin without a cat. A field without a medium. Curious indeed. You'd need the imagination of Lewis Carroll or Albert Einstein to conceive of such a thing.

Is it really possible? The math of Einstein's relativity formulas shows that it is. It has become quite common, in recent decades, for math to teach physicists that something is logically consistent even though it seems as if it shouldn't be.

Ten years later, however, Einstein reconsidered his views. By then he'd realized that empty space is the gravitational field's medium and that it has all the quirks of the luminiferous aether, including amotionality and the nightmare property. Empty space is to the gravitational field as the luminiferous aether is to the electromagnetic field; they're equally weird.

That left Einstein with three possibilities:

1. Empty space exists, but the luminiferous aether doesn't.
2. Both empty space and the luminiferous aether exist.
3. Neither empty space nor the luminiferous aether exists.

The first choice seemed inelegant and asymmetric. He'd found formulas for the gravitational and electromagnetic fields that fit together in a simple way, preserving the same cosmic speed limit. Why should one field's medium exist while the other's does not?

The third possibility was clearly unacceptable. If empty space, the place we live in, isn't real, then what is?

This led Einstein to conclude that the luminiferous aether probably does exist after all. If so, it gives us a second cosmic triad: the luminiferous aether, the electromagnetic field, and electromagnetic waves.

Many physicists still prefer Einstein's original view that the electromagnetic field is a Cheshire Cat field. Experimentally speaking, the question is open and perhaps unanswerable. But if, like Einstein, we adopt the second option, then the luminiferous aether, rather than resembling an ordinary medium, must be more like empty space: transparent, permeable, everywhere, amotional, and nightmarish. Despite the resistance of our common sense, we have to accept the possibility of such media because, as we've seen, empty space is one of them.

Around 1920, Einstein became interested in the work of Theodor Kaluza, extended a few years later by Oskar Klein. Kaluza considered the possibility that space has four dimensions rather than three (i.e., that space and time together make up five dimensions rather than Einstein's famous four). In Klein's terms, we would imagine that one of these dimensions is so small that we don't notice it, somewhat as a child looking at a sheet of paper might not realize that it has a thickness. We encountered this already in Fig. 33 (p. 178), when I pointed out that a paper-thin aluminum sheet has not only an obvious bending field but also less obvious leaning and compression fields. Kaluza and Klein collectively noticed something similar about adding one short extra dimension to empty space. To those of us too large to detect that extra dimension directly and aware of only our usual three space dimensions, Einstein's gravitational field in the full four space dimensions would appear as multiple fields: a gravitational field (bending), an electromagnetic field (leaning), and one more field (compression) often called the *radion*. The latter is a topic for a different book.[17]

Kaluza's and Klein's mathematical studies suggested that the electromagnetic field might not have its own distinct medium—that the luminiferous aether might not be separate from empty space. Instead, much as mass density and magnetization are both properties of iron, the electromagnetic field and the gravitational field might both be properties of the *same* medium— an empty space with more structure than meets the eye.

Einstein, finding this idea compelling, spent much of his later life trying to develop it as part of his failed attempt to discover a successor to quantum physics. Today, the notion that the elementary fields of nature might all arise as properties of a single medium (or a very small number of media) remains popular. For instance, it naturally happens in string theory, albeit in a more elaborate fashion than I've described here.

Yet there's no experimental evidence for this idea, and it still might be the case that the electromagnetic field is a Cheshire Cat field, with no medium. In fact, all the elementary fields might be Cheshire Cat fields, with the gravitational field the lone exception. Perhaps Einstein was right the first time.

Or perhaps he was wrong both times. There's still the third option. Might the bizarre features of the universe, its amotionality and its nightmare property and so on, be clues that empty space doesn't actually exist, either?

This may well sound ludicrous. (It did to me when I first encountered the idea.) But remember the story of color—how real it seems to be, even though it's largely manufactured in the human brain. *What do you mean, the sky isn't blue? And there's no such thing as white or pink?* The empty space between objects seems real. Maybe it's not. Maybe it's a way of thinking, an unnecessary one, and perhaps even an obstruction to seeing the universe for what it actually is. Physicists do study imaginary universes in which the space we think of as real is only an optional crutch for understanding the world, a crutch that, while sometimes convenient and sometimes not, is never absolutely required.[18]

The remainder of this book, a story of fields found throughout the cosmos, implicitly rests on the assumption that empty space exists. But keep in mind that it might not. If it doesn't, the story I'm telling here will have to be translated someday into a different conceptual language. Don't be too surprised if and when that happens. If there's anything I've learned in my career, it's that quantum physics in the context of space and time offers the universe far more alternatives than Einstein and his contemporaries could ever have imagined. Abandoning space and perhaps even time might someday help us understand why the cosmos is so fantastically, mind-bogglingly odd.

But let's set that issue aside for this book and return to Einstein's line of thinking, where various elementary fields may represent properties of a single medium. Are there any clues that might support such an idea? Here's one: the cosmic speed limit. Perhaps the reason that waves in all cosmic fields satisfy the same speed limit is that they share a common origin.

Seen from this perspective, the significance of Kaluza's realization about gravity and electromagnetism looms a bit larger. Not only did he show that the electromagnetic field and gravitational field might emerge from just one medium—an enhanced version of empty space—he also proved that their waves were destined to travel at equal speeds. Klein, through further analysis of Kaluza's ideas, learned that this is not the full story. There are other fields (now called Kaluza-Klein modes[19]) that would arise if space had additional microscopic dimensions. Although waves in those fields would not travel at the speed c, they would all respect the same cosmic speed limit, just as the elementary fields in our cosmos do.

Could it be, then, that all the fields of nature spring from a shared medium, perhaps a generalized notion of empty space with a built-in fundamental speed limit? This would be elegant by the standards of scientists. So far, there is no experimental evidence for or against this idea, and no one with Einstein's brilliance has been able to argue convincingly that it must be true.

Yet it's possible that the idea is misguided. First, even when multiple fields derive from just one medium, this does not guarantee that all the fields will share the same speed or speed limit. When an earthquake occurs, there are several types of seismic waves with different speeds, even in the same type of rock. Vibrations in the bending field of a piece of iron don't move at the same speed as do its spin waves. Second, the cosmic speed limit may merely tell us about relativity itself, revealing nothing about the fields' origins. As we saw, Galileo's principle compels light and gravitational waves to be nightmarish and requires that their media be amotional, necessitating a major adjustment to the workings of time and space. This adjustment leaves no further wiggle room. If the waves of any elementary field traveled at a constant speed in empty space that was different from c, or respected some other limit than c, or respected no speed limit at all, then their medium could not be amotional. Those waves would then provide illegal information that we could use to measure our motion in an isolated bubble, in defiance of the relativity principle.

We are left with no firm conclusion. Have we learned anything at all? It seems that Galileo's principle of relativity and Einstein's cosmic speed limit are firm. There are hints of a universal but amotional medium, some sort of enhanced, generalized conception of empty space. But beyond that, we have little to go on. Our intuition isn't much help. After all, if we didn't know that an amotional medium is possible—if experiments hadn't observed empty space acting like an impossible sea—we wouldn't have believed it.

Medium	Field	Wave
Air	Wind	Ordinary sound
Water	Pressure	Pressure waves
Iron	Mass density Magnetization	Density waves Spin waves
Aluminum sheet	Bending Compression Leaning	Transverse waves Density waves Shear waves
Empty space	Gravitational	Gravitational waves
Luminiferous aether (?)	Electromagnetic	Electromagnetic waves
Higgsiferous aether (?)	Higgs	Higgs waves
Imaginary empty space with one small extra dimension	Gravitational Electromagnetic Radion Kaluza-Klein modes	Gravitational waves Electromagnetic waves Radion waves Kaluza-Klein waves
Generalized empty space (??) [Our universe?]	All elementary fields	All elementary waves

Table 3: Various media, with one or more of their fields and corresponding waves. A question mark indicates media that might or might not exist, even though their fields do. The imaginary empty space with one extra dimension is an example of how a wide variety of elementary fields might arise from one everywhere-medium. Could our universe be similar?

15

Elementary Fields
A Second, Humble Look

The last chapter has led us deep into an intractable morass, thick with confusing questions that trouble even professionals. We'll escape this swamp in a most humbling way: by declaring temporary defeat and backing up.

Fortunately for professionals and nonexperts alike, we can make a surprising amount of progress by taking a different route, keeping to the field-centric perspective for the rest of this book. Though the notion of the universe-as-medium will not entirely retreat from view, we will focus on the elementary fields. There the mysteries, though deep, are more manageable, and many lessons can be learned.

Despite the numerous gaps in our basic conceptual understanding, the study of the elementary fields has been a remarkable success. Over the past century, scientists have discovered formulas that are spectacularly effective at describing how these fields behave and interact. Astonishingly, as of 2023, there is not a single phenomenon observed in an experiment that directly and uncontroversially contradicts these formulas. Among all human inventions and discoveries, they're the most accurate.

In short, we have a remarkably clear (if incomplete) picture of what the known elementary fields *do*. Despite this, we have barely any concept of what they *are*—assuming that's even a question we should be trying to answer.

In this context, this chapter's goal is to highlight the essential aspects of elementary fields that we will need for the remainder of the book. There follows a summary that you can refer back to as you continue to later chapters.

Let's first establish some common language. What we refer to as *calm* for the wind field and *unmagnetized* for the magnetization field of iron are very similar concepts; in both cases, a measurement of the field (the wind speed and direction anywhere within the atmosphere or the magnetization field anywhere within the iron) would yield zero. Similarly, when saying "the Higgs field is switched off," I mean that if you measured it anywhere, you'd find it is zero. As common terminology, we'll say that a field with these characteristics has an *average value of zero*.[1]

I have used the word *average* for a reason, though I'll sometimes drop it when the context is clear. Even on a calm day, the wind's speed won't be exactly zero everywhere all day long; there are always little light breaths of wind, eddies and vortices, and sound waves. Similarly, the magnetization field of unmagnetized iron is zero on average but not atom by atom, as shown in the middle section of Fig. 32 (p. 175). These little variations over time and from place to place will usually not concern us; we will focus on the average value of a field as a way to characterize the field's general behavior.

Similarly, what we refer to as a *steady breeze* for the wind field, *magnetized* for the magnetization field, and *switched on* for the Higgs field are three manifestations of a field having a *nonzero average value*. Any measurement of the value of the field, averaged over a certain region of time and space, gives the same nonzero answer. I will also refer to this as a *constant, uniform field*.

I introduced the wind field at the start of this section because it's a typical ordinary field. Found throughout its medium, it's dynamic, meaning that its behavior at one place and time affects its behavior nearby at a later time. Though often turbulent and complicated, its activity is sometimes very simple: it may be calm, steady, or, when sound waves pass by, wavy. These features are worth noting because they are also common among elementary fields.

On a calm day, an English speaker might say either "There's no wind" or "The wind is calm." They mean the same thing, and yet conceptually there is a slight difference. The first implies that *the wind doesn't exist* on a calm day. The second implies that *the wind does exist* but is just inactive.

In physics dialect, we would use only the latter phrasing. For us, the wind exists and can be measured anywhere, at any time, within the Earth's atmosphere. The measurement might reveal a calm wind, but that doesn't mean the wind has ceased to *be*. For physicists, a field always exists anywhere and anytime within its medium.

A field does not exist outside its medium, however. There is no such thing as wind beyond the atmosphere, where there are no gases that can flow. Any ordinary field exists only within the finite region occupied by its medium.

A cosmic field is different, as it has either an everywhere-medium or (if it is a Cheshire Cat field) no need for any medium. It exists everywhere throughout the universe, both in empty space and inside all material objects.

Across most parts of the universe, particularly in deep space, the average value of almost every known elementary field is zero. The Higgs field is the notable exception. Near Earth, more fields are nonzero. If we look immediately around us, we find that we are surrounded by a uniform gravitational field, which assures that any objects dropped near us accelerate in the same direction at the same rate, as well as a uniform magnetic field, which assures that all nearby compasses point in the same direction.[2]

From these examples, we learn that when a certain field has a nonzero average value, objects that interact with it will behave differently than they would otherwise. That's clear enough for the wind field: a steady breeze extends all flags and bends all trees. The nonzero gravitational field around us causes objects to fall; the nonzero magnetic field aligns our compasses. And the Higgs field's nonzero average value assures that certain elementary particles have rest masses. None of these would happen if these fields' average values were zero.

Sound is a wave in the wind, and similar waves exist for all the elementary fields of the universe. This is no surprise, as waves are so generic; if you disturb an ordinary medium or field in any sudden way, you will generally cause waves to form. Many cosmic fields were discovered through their wavy behavior. Some of their waves can travel all the way across the universe, like light, though possibly more slowly. Others dissipate in moments, becoming waves of other fields in much the way that waves on a guitar string dissipate into sound waves. Still others are trapped inside objects, as gluons are trapped inside protons and neutrons; they can travel across the universe, but only when their prisons do.

The presence of simple waves does not change a field's average value, because the back-and-forth variation created by such a wave averages out over time. In fact, a field's waves may be found doing one thing while the field's average value is doing something entirely unrelated. To translate a point I made earlier into field-centric language, we can hear sound waves arriving from the north even when there's a steady breeze from the east. Similarly, the fact that our local magnetic field points toward the Earth's pole does not prevent light waves from reaching us from all directions.

All fields can have traveling waves; localized disturbances with enough energy can always make them. Standing waves, though impossible (in empty space) for the electromagnetic and gravitational fields, are common to most of the universe's fields. They will play a central role in future chapters.

Up to this point, I have been revisiting and restating facts that we have already encountered. Now we need to turn to something very different—a profound and nonobvious consequence of relativity, quantum physics, and the cosmic speed limit.

It turns out that cosmic fields fall into two sharp categories: *bosonic* and *fermionic*. This division has many implications, which we'll trace throughout the rest of this book. *Bosonic* fields are named after the Indian physicist Satyendra Bose, who in the 1920s greatly impressed and inspired Einstein with his work.[3] Among elementary fields, bosonic fields include the gravitational, electromagnetic, and Higgs fields; others are called the *gluon field*, the *W field*, and the *Z field*. Fermionic fields are named after Enrico Fermi, one of the widest-ranging physicists of the twentieth century.[4] Among the fermionic elementary fields are six types of quark fields (named up, down, strange, charm, bottom, and top), three neutrino fields, and three electron-like fields (the electron field, muon field, and tau field).

The key thing to remember is that bosonic fields are free spirits, while fermionic fields are tightly restrained. The average value of a bosonic field can be zero or nonzero, and if nonzero can potentially be of any size. More generally, a bosonic field's value at *any* specific location and time can be zero, small, or large. This is true of the wind, of course. Furthermore, the magnetic field around the Earth has a larger value close to the poles than it does near the equator, while the gravitational field is largest near the Earth's surface and becomes smaller as one moves into deep space.

Finally, the amplitudes of waves in bosonic fields can be large, as in a laser. But they can also be ultra-microscopically small, as for single photons.

None of this variety is available to fermionic fields. Their waves can have only a microscopic amplitude, and their average values can only be zero. Even their values at specific locations can only be microscopic.

These limitations on fermionic fields explain why only bosonic fields can serve as long-distance intermediaries, inducing forces across gaps between separated objects. It's not an accident that the forces we've spoken about so far involve the gravitational and electromagnetic fields, both of them bosonic. In accordance with its name, it's the gluon field, another bosonic field, that generates the forces that hold the proton together. The fermionic electron, quark, and neutrino fields cannot serve as direct intermediaries between objects.[5]

All wavy phenomena that we can discern in ordinary circumstances, such as waves of water, sound, and light, are bosonic waves. Sound waves can be almost imperceptible, but they can also be deafening; a laser can be blinding or dim. Something analogous is true of gravitational waves, and even (though impractical and brief) Higgs waves. By contrast, the intensity of a fermionic field's wave cannot be adjusted; a bright laser-like wave is impossible. Fermionic waves are so feeble that we never notice them around us and have no intuition for them. Nevertheless, they are just as essential to our lives as bosonic waves, as we'll see.[6]

15.1 Fields: A Summary

This is a good moment to take a breather. Before we move on, I'll bring the highlights of this section of the book together in a short summary, followed by two tables that you may find useful for later reference. Then, equipped with the novel ideas from these chapters, we will turn to quantum physics, where we will learn the true nature of particles, rest mass, and the Higgs field.

From ordinary fields such as the wind field of the atmosphere and the magnetization field of a piece of iron, we learn that

- An ordinary field is a property of a medium, measurable at any location within the medium.
- The field can vary from place to place and over time, and what the field is doing now will influence its future some distance away.
- The field's impact on the world is sufficiently predictable that its value at any place and time can be measured by scientific instruments.
- Any single type of ordinary field can arise, potentially, from many different ordinary media.
- An ordinary medium with many properties can be the host of many different ordinary fields.
- A field may reveal a property of a medium that may not be obvious to the naked eye or to simple experiments.
- An ordinary medium inevitably obscures the principle of relativity but does not ruin it; although one can measure one's motion relative to it and it can exert drag that depends on one's motion, it can be excluded from an isolated bubble where the principle of relativity must hold.

Cosmic fields bear some resemblance to ordinary fields but show essential differences, too.

- A cosmic field exists everywhere in the universe and may be a property of an everywhere-medium.
- Cosmic fields may be composite (made of other fields) or elementary.
- Unlike an ordinary medium, an everywhere-medium cannot be escaped; it is found even inside isolated bubbles.

- To respect the principle of relativity, any everywhere-medium, including empty space, must be amotional—speed relative to it cannot be measured or even defined. This is a property no ordinary medium can have.
- Any medium for an elementary field must be transparent and permeable.
- Any waves that travel at the cosmic speed limit must be nightmarish—impossible to catch up with and, if they are chasing us, impossible to escape.
- Any waves in an amotional medium must respect the cosmic speed limit from all observers' perspectives.

Understanding these everywhere-media so little, and knowing no examples except empty space itself, physicists mainly resort to the field-centric perspective, in which waves are thought of as ripples in fields rather than in some unknown medium.

- All the known elementary fields have traveling waves; some have standing waves, too.
- Both bosonic and fermionic fields can have an average value of zero, but only bosonic fields can have a nonzero average value.
- Only bosonic fields can have waves of large amplitude; the waves in fermionic fields are always microscopic.
- Only bosonic fields can serve as direct intermediaries between objects, leading to forces across gaps.

Field	Ordinary	Cosmic	
A property of	an ordinary medium	(perhaps) an amotional everywhere-medium	
It exists	within its medium	everywhere across the universe	
		Composite	**Elementary**
Made of other fields?	Yes	Yes	No

Table 4: Comparing ordinary and cosmic fields, which can be composite or elementary.

Elementary Fields	Bosonic Field	Fermionic Field
Average value	May be zero or nonzero	Always zero
Value at specific location and time	Unrestricted	Zero or microscopic
Waves' amplitude	Unrestricted	Microscopic only
Forces between distant objects?	Potentially	No
The known elementary fields	Gravitational field, electromagnetic field, gluon, *W*, *Z* fields, Higgs field	3 neutrino fields, 3 electron-like fields, 6 quark fields

Table 5: The features of bosonic and fermionic fields and the separation of the known elementary fields into these two categories. The electron-like fields are named electron, muon, and tau, while the six quark fields are named up, down, strange, charm, bottom, and top. Naming conventions for neutrinos are currently in flux.

QUANTUM

We live in a quantum universe. Quantum physics drives the modern economy: lasers and LEDs operate on quantum principles, as do the transistors in our phones and computers. Quantum cryptography, quantum computing, and other advanced quantum technologies lie in the relatively near future. But the whole universe runs on quantum physics, too. As was realized in the 1920s, atoms would collapse without it. Soon after, it was understood that what we've been calling *particles*—photons, electrons, quarks, and so forth—arise directly from the quantum physics of fields.

Scientists came to understand the relationship between particles and fields by combining Einstein's relativity and quantum physics. In the process, they developed what we now call *quantum field theory*.

"But is it all just theory, then?" asked an acquaintance over lunch. "Not fact?"

"Ah!" I replied. "That's a point of common confusion. For physicists and most other scientists, *theory* means something different from what it means in standard English!

"Scientists usually aren't distinguishing *theory* from *fact*. Instead, *theory* refers to a set of mathematical formulas, accompanied by a set of underlying concepts that explain how to use the formulas to predict nature's behavior. When people talk about Einstein's 'theory of relativity,' that's what they mean: the equations and concepts that Einstein introduced. Even when convincing experimental evidence piles up in favor of a theory, so that we'd start calling it a fact in English, scientists continue to refer to it as a theory."

"So is that true of 'string theory,' too? It's math and concepts?" he asked.

"Yes, string theory is also a set of formulas resting on a foundation of concepts," I affirmed, "but its connection with the real world is far more limited. Einstein's ideas about relativity have been confirmed in scientific experiments and in technology; his formulas do an excellent job of describing how nature works. String theory, when used in an attempt to explain the fundamental laws of nature, hasn't been and can't easily be tested experimentally, and so it remains speculative. So string theory is a theory *both* in the physics sense *and* in the English sense, whereas Einstein's relativity, a theory in the physics sense, is essentially a fact in English."

"Oh, only 'essentially' a fact?" he chuckled. "Hedging your bets, are you?"

Well, I was being judiciously precise. No theory—no set of formulas—can ever reach the level of English fact; at best, it must stop just short. Just as it would be more honest to refer to electrons as up-to-now-apparently-elementary, the theory of relativity should be viewed as up-to-now-apparently-fact. There's no evidence against it today, but more powerful technology might someday reveal weak points. In fact, that's what happened to Newton's laws, which had seemed perfect for centuries.

Experimental results are much closer to English facts. In an experiment, you set up a controlled situation and watch to see what happens. The result, which teaches you something concrete about how nature operates, becomes part of your permanent knowledge base. That knowledge base isn't subject to revision merely because a new and better theory has come along. Quite the contrary; the results of experiments hold their ground. If a newly proposed theory is inconsistent with a host of previous experiments, then it is inconsistent with nature and must be rejected.

Theoretical predictions that match experimental data survive, too, though in a more subtle way. Even after Einstein revised Newton's formulas on which two centuries of technology had relied, there was no upheaval in the economy. Bridges and skyscrapers that were built using Newton's laws remained upright; engine manufacturers who relied on Newtonian predictions weren't suddenly obsolete. Only in experiments unknown to the nineteenth century did Einstein's new laws show clear superiority to those of Newton. Where Newtonian predictions had already matched experiment, Einstein's predictions were almost identical, with tiny, irrelevant differences, though they were obtained using methods and concepts quite different from those of Newtonian theory.

More generally, any new theory's predictions need to agree well with the successful predictions of the theory it seeks to supplant. However, the details of how the math and concepts of the new theory generate those predictions may differ sharply from those of its predecessor. Herein lies the important lesson: experimental results and the predictions that have matched them are stable, but math and concepts are always potentially subject to revision. Any set of math formulas—any theory—can never be more than up-to-now-apparently-fact. We can never preclude the possibility that future experiments will someday reveal it as only approximately true.

Prior to quantum physics, nineteenth-century field theory worked well for sound, ocean waves, and even light. It was in good accord with common sense, too. But in the early twentieth century, experiments showed that it is only approximately true. Quantum field theory is its modernized version, cruel to common sense but far more successful in its match to experiments, and as close to a fact as any theory in science.

The most important prediction of quantum field theory is that a quantum field's waves are made from "particles." The classic example is that of the electromagnetic field: its waves are made from photons.

The light wave from a bright laser is a steady, simple traveling ripple in the electromagnetic field; it has a definite frequency and a large amplitude. A photon is much the same except that its amplitude is tiny, almost imperceptible. The laser light, or indeed any bright light such as that produced by the Sun or a lightbulb, is built from immense numbers of these faint little ripples all traveling together.

We'll explore photons in more detail in the coming chapter. But let me first dispel a common misconception. A photon is not a wave *in the sense of standard English*; it is not a single wave crest. You don't count photons by counting wave crests. Instead, a photon is a wave in the sense of physics dialect: a ripple that may have many crests and troughs, also known in English as a wave train; see Fig. 22 (p. 141). As we'll see, you count photons by looking at other properties of a wave, such as its energy.[1]

But if photons are ripples with crests and troughs, moving along at the cosmic speed limit just like any light wave, in what sense are they *particles*? Nothing I've said so far makes this obvious. Our experiences with familiar waves give us no insight into this relationship between waves and particles; although our visual systems rely upon it, it's something none of us has ever seen. To bring this mysterious feature of the cosmos to light is the goal of the next chapter.

16

The Quantum and the Particle

As I've hinted, even the word *particle* is a false friend. Despite how electrons and photons and quarks are often drawn, they aren't much like dust particles or grains of sand. They're much more like waves. But to understand how this is possible, we need to go back to Chapters 10 and 11 on vibrations and waves, review their lessons, and identify one little mistake.

The first of these lessons was that any simple vibration has two main properties: an *amplitude* (how far does the object wiggle?) and a *frequency* (how often?). Allowed to move freely, many objects vibrate with a preferred frequency, called the *resonant frequency*. The stability and reliability of this frequency assure that pendulum clocks keep good time and that a guitar string makes a pure and predictable note.

Simple waves, such as that shown in Fig. 22 (p. 141), are themselves a form of vibration, with an amplitude and a frequency. Traveling waves, whose crests and troughs move from one place to another, have speed and a direction; they can be made at any frequency. But simple standing waves have crests and troughs that don't go anywhere. The simplest of these standing waves (see Fig. 25 on p. 145, upper left) is what we find on a plucked guitar string, with just one crest or trough. This special wave, like any resonant vibration, vibrates at a unique frequency.

Changing a vibration's amplitude is easy; a guitar player can choose the amplitude by deciding how hard to pluck a string. Changing a resonant vibration's frequency is more challenging; the player can shorten a guitar string, tighten or loosen it, or change its environment.

These central lessons were familiar in the nineteenth century; many were already known in ancient Greece. Yet one of these lessons is a tiny bit wrong.

What's wrong is that we cannot choose the amplitude of a vibration as freely as I've implied. For any type of vibration or wave, there is a smallest possible amplitude that it can have. If you try to make the amplitude smaller, the vibration or wave simply won't occur.[1]

Common sense yet again resists this strange assertion, claiming that it is easily refuted. Take a pendulum, choose an amplitude for it, and watch it swing. Now, make the amplitude half as big and try again. Then half as big again. Then half as big again. You can continue this process indefinitely until the amplitude is as small as you could ever want. There's no minimum amplitude, because it can always be halved. The logic is simple. What could possibly be wrong with it?

If you really tried it, you would eventually find, after many steps, that your procedure would fail. You would be unable to reduce the amplitude by half at each step. Instead, the pendulum would vibrate with amplitudes slightly different from the ones you wanted. Finally, it would reach its minimum amplitude, and your only options after that would be to leave it as it is or to shut off the vibration altogether.[2]

Although quantum physics emerged over a century ago, and modern technology rests upon it, scientists and philosophers are still arguing over how to think about it. Fortunately, if our goal is to understand motion, mass, relativity, and fields, we don't need a thorough introduction to quantum physics. We need only focus on this one essential element, so alien to our ordinary experience and yet so important: that there exists such a thing as the faintest possible tremor. This minimal vibration, with the smallest possible amplitude, is what is called a *quantum*.

We can't intuit quantum physics from daily life. Any ordinary object we might encounter, large enough to be seen or touched, is shaking with far more than one quantum's worth of vibration. If we stumbled across an object vibrating with just a single quantum, it would appear completely still; worse, just the act of shining light on it to make it visible would increase its vibration above its minimum.

For this reason, I have no do-it-yourself experiment to suggest this time. To prove to yourself that the world is quantum requires either expensive equipment or a conceptual argument based on experimental data. That's why no one recognized this remarkable property of nature until the twentieth century.

The principle of a minimum possible amplitude applies to waves, too. This includes traveling ripples in a metal table, standing waves on a guitar string, and electromagnetic waves. Our eyes are built to work on quantum physics; they can actually detect a single quantum of light. But what do we really mean by a quantum of a wave? What exactly is a photon?

Perhaps you have a lightbulb in your house that works on a dimmer switch: a knob or slide that adjusts to make the bulb gradually brighter or dimmer. Let's imagine turning such a bulb on full, and then, in analogy to what we tried with a pendulum, let's use the dimmer switch to make the light half as bright, and again half as bright, and half as bright again. With ever smaller adjustments to the switch, we can make the light as dim as we want, reducing the light wave to whatever amplitude we choose. To be sure, this would take a more sophisticated switch than we actually have in our homes. But in principle, couldn't we do this?

No. It won't work.

As we turn the switch down again and again, we will eventually notice that the light is no longer of constant brightness; it will start, subtly, to waver. After further steps, the bulb's wavering will turn into an unmistakable flicker. And once the dimmer switch is low enough, the bulb will remain off most of the time, emitting only infrequent flashes at random intervals.

Nature outwits us. Instead of the light diminishing to a continuous but extremely faint glow, as we would have anticipated, it will instead become a discontinuous dribble of rare flashes, each of a low intensity. If we replaced the lightbulb, which emits light at many frequencies, with a laser beam,

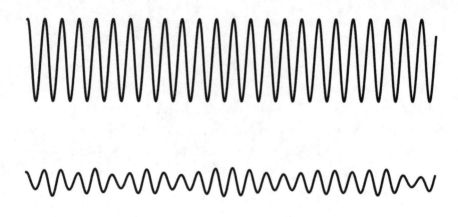

Figure 37: Light waves from a laser at high, medium, and low intensity, traveling to the right; schematic, and not drawn to scale. (Top) A steady, simple wave. (Middle) The light becomes unsteady. (Bottom) The light occasionally flashes at random times; each flash, a brief ripple, is a photon.

whose light has a single, pure frequency, there'd be only one important difference: each of the occasional flashes would have identical brightness, as illustrated in Fig. 37.

This unexpected observation reveals that for light, there is a dimmest possible flash. Such a flash is an electromagnetic wave with the smallest permissible amplitude—the electromagnetic field's gentlest ripple. The ripple can take any shape; it may have many crests and troughs or just a few. Its frequency may lie in the visible range or anywhere else on the electromagnetic spectrum, from radio waves to gamma rays. But its overall tiny amplitude cannot be tinier.

This dimmest possible flash is what we call a *photon*, or a *quantum of a light wave*, or, more simply, a *quantum of light*. It is a minimal ripple, as delicate an undulation as you can create in the electromagnetic field. If you try to make less, you'll make nothing at all.

One of my adult students knit her brows. "So despite the name," she complained, "a photon's really a wave."

I nodded. "A photon can have an amplitude and a frequency just like any simple wave, so yes, in the physics sense of the word, it's a wave."

"Ugh…" she groaned in disgust. "Isn't there a better name than *particle* for this thing?"

I admit we've yet again made a mess of language. I'm not really a "particle" physicist.

There was an attempt, back in the 1920s, to introduce a new and more descriptive word. The idea was to refer to a quantum of a wave as a *wavicle*, making it clear that it is somehow in between particle and wave.[3] The word is rarely used today, but it seems to be making a comeback. I like it because I think it captures what's different about a photon from what we might draw or imagine when we hear the word *particle*. I'll use the term for the remainder of the book.

The reason that wavicles such as photons have been called *particles*, historically, is that many of their properties are particle-like. A wavicle is as small in amplitude as it can possibly be, so you can't divide it in half, or into thirds, or into any sorts of pieces. If you did, each piece would have a smaller amplitude than is allowed. *So a wavicle behaves as a fundamentally and absolutely indivisible entity.* It travels as a unit. You can throw one, or catch one, or bounce one off a wall; an atom can emit one or absorb one. But you'll never find an atom emitting a third of a wavicle, nor can you hold 2.4597 wavicles in your hand. Quanta in general, and wavicles in particular, are like people. You can have 1, 2, 7, none, 465—but not fractions. Nothing you do in the universe can ever make a fraction of a photon.[4]

We might say that a photon, and a wavicle more generally, is a *particulate wave*—a wave that, much like our naive idea of an elementary particle, cannot be disassembled into smaller pieces. Almost by definition, it has extraordinarily strange properties. Like any water or sound wave, a photon can spread out, even across a large room. Yet if it is absorbed into the walls of the room, a single atom, located at one microscopic spot somewhere on the room's walls, will take it in wholesale, swallowing it in one gulp. First the photon is widely dispersed, and then, somehow, it isn't. If this were a book focused on quantum physics, we'd spend multiple chapters thinking about that one sentence. But we're just going to smile and move on. Wavicles seem very strange, but such details are for another time.

This would be fascinating enough even if it held only for photons. But electrons are wavicles, too. Just as a photon is a quantum of a wave in the electromagnetic field, an electron is a quantum of a wave in another field called, simply, the *electron field*. Do be careful not to conflate the electron field with the more familiar electric field, as they are completely different. For one thing, the electron field is fermionic, while the electric field, a part of the electromagnetic field, is bosonic; see Chapter 15 and Table 5.[5]

In fact, all types of elementary particles, from neutrinos to Higgs bosons, are wavicles of fields. A down quark is a wavicle in the down quark field, an up quark is a wavicle in the up quark field, and so on. Each of the three types of neutrinos is a wavicle in one of the three neutrino fields.

Among the bosonic fields, a W boson is a wavicle of the W field, a Z boson is a wavicle of the Z field, and a gluon is a wavicle of the gluon field.[6] Last but not least, a Higgs boson is a wavicle, too, the gentlest possible ripple in the Higgs field. That's why the discovery of the former proved to scientists that the latter exists as well.

Each of these elementary fields may or may not be a property of an amotional everywhere-medium, for the reasons discussed in Chapters 14 and 15. But nothing is known about those media, and none have been named. Just as for light, and for the same reasons, scientists stick closely to the field-centric perspective for all these fields and their wavicles.

A side note about antiparticles, which I discussed briefly in the context of the proton, back in Chapter 6.4. Concerning anti-quarks, you might wonder whether there are anti-quark fields as well as quark fields or whether there are even anti-fields. This is partly a matter of convention. But I think the easiest way to think about it is this: some fields have two types of wavicles, while others have only one. In the first case, the two types of wavicles are each other's antiparticles, while in the second, the lone wavicle is its own antiparticle.[7] Up quarks and up anti-quarks are in the first category; both are wavicles of the up quark field. Similarly, electrons and positrons are both wavicles of the electron field. But the electromagnetic field is in the second category; it has only one type of wavicle, and so the photon is its own antiparticle.

And one more side note, to head off a common and problematic confusion for those who've already read about, heard about, or even learned a little quantum physics. (Other readers can safely skip this paragraph.) It is easy to become confused about wavicles because a second wavy concept often arises in discussions of quantum physics. Wavicles such as electrons and photons are physical objects that carry energy; they exist within and move across the same empty space that you and I live in. An unrelated wave is called the *Schrödinger wave function*; it does not exist within the space that you and I live in, does not carry energy, and is not a physical object at all. Instead, *it is a wave that travels in the abstract space of all possibilities for the wavicles and fields that it describes.* This wave function is a mathematical tool used to calculate the probabilities for what physical objects (including wavicles and fields) may do. Wavicles, as physical objects, can have observable consequences—a wavicle can enter your body and cause damage to one of your cells, for instance—while a wave function can do no such thing. We will not encounter wave functions again in this book.[8]

Now I want to return to a point from Chapter 11. There I emphasized that waves bear a remarkable similarity to ordinary objects; both a ball and a wave can carry energy from one place to another and knock a cup off a table. At the time, I made it seem surprising that these two disparate things could have so much in common. I described a wave as a transient ripple in a medium, obviously very different from a material object like a ball.

But this conception of what a wave can be, derived from ordinary experience, proves far too narrow. Wavicles violate our common sense about waves. Unlike the waves we see around us, they are particulate and can be used as the building blocks of more complex objects. They can bind together under the influence of forces. Some can last indefinitely. Nonetheless, they remain waves, with frequency and amplitude.

Since a ball is made entirely from wavicles of fields, the similarity of a ball to a wave on a rope isn't a surprise after all. Ordinary material consists of a huge number of tiny ripples all traveling together, carrying energy from one place to another.

To make this explicit, let's revisit the comparison. We'll do it in reverse order, looking at the waves from the cup's point of view (Fig. 38).

Initially, a rope is sitting in the vicinity of the cup, doing nothing. A wave on the rope approaches and interacts with the cup, knocking it over. The wave continues on, leaving the rope behind in its original, waveless state.

Now, the ball—but here I'll use the field-centric perspective instead of the medium-centric one. Initially, the fields of the universe are sitting in the vicinity of the cup, doing nothing. Then a ball-shaped crowd of wavicles of those fields—electrons, quarks, anti-quarks, and gluons—approaches the cup. Some of the wavicles at the edge of the ball interact with the cup, knocking it over. The ball of wavicles then continues on, leaving the fields behind in their original, waveless state.

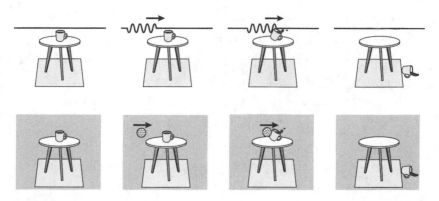

Figure 38: (Top) A rope stretched near a table; a wave approaches, strikes a cup, and travels on, leaving the rope behind. (Bottom) The universe's fields (shaded gray); wavicles in a ball approach, strike a cup, and travel on, leaving the fields behind.

The analogy might seem imperfect: the ball survives the collision, while the wave on the rope disappears. But this is a detail. (It has to do with dissipation: the rope's wave can quickly dissipate away, while the tiny electron and quark wavicles in the ball cannot.) It is irrelevant to the analogy's most salient point: on a rope or within elementary fields, traveling waves can enter from afar, upset the cup, and depart. As they vacate the area, the waves leave the rope and the elementary fields behind, as placid as they were before the waves arrived.

In the same way, the air (and the wind field) are only temporarily disturbed by the sound waves from a guitar string. Once the sound dies away, they are left in place, just as they were. If an outdoor light is flicked on and off on a dark night, as in Fig. 39, the initially zero-valued electromagnetic field vibrates for a few moments as the photons pass, and then returns to zero. And if an electron, a hawk, or a star travels by, it disturbs the fields of the cosmos only briefly. Ever-present and woven into the substance of the universe, these fields act as the supporting structure within which we and all other things take form.

Figure 39: (Top) Before a lightbulb flashes, the electromagnetic field's value is zero. (Middle) Once it flashes, waves of light move through the electromagnetic field. (Bottom) After they pass, the electromagnetic field is again zero.

$=$ 17 $=$

The Mass of a Wavicle

I don't know why the word *wavicle* didn't catch on a century ago. Probably it's just another historical accident. I suspect that Niels Bohr, a dominant figure in quantum physics in its early years before quantum field theory came into its own, just didn't like it; it didn't really fit into his way of thinking.

Words carry history with them. Scientists first encountered electrons in the 1870s as mysterious "cathode ray" beams that traveled in straight lines. They eventually discovered that the beams were made of identical electrified objects that come in ones and twos and threes but never in fractions. They called those little units of electricity *electrons*. Next they realized that each type of atom has a certain number of electrons on its outer edges. They initially envisioned electrons as points or balls, like dust particles shrunk down to microscopic size, moving around in the way that dust particles do.

By the 1920s, when it was finally recognized that electrons don't have trajectories like balls and inhabit atoms more by surrounding than by orbiting, the concept of an electron as a particle-like object was locked into scientific language. This early erroneous conception of electrons survived into physics dialect and its translation into English. It also endures pictorially in the atomic cartoon, where it continues to mislead an entire species. A false picture is worth a thousand false words.

Meanwhile, Einstein's idea that electromagnetic waves might be made from tiny packets was more an inspired guess than a crisp insight; even in

1905, Einstein himself was well aware that it had serious conceptual problems. At the time, he had no idea that all vibrations, not just light waves, come in quanta. Only in the 1920s did experiments reveal that his quanta of light were "particles" in the same sense as electrons.

During the following decades and culminating in the 1970s, it became clear that the universe can be understood in terms of elementary fields of nature. Particle physicists became wavicle physicists, albeit without a name change. Ever since, wavicle experts have spent much of their time developing, studying, and interpreting the formulas of quantum field theory, which serve to describe the behavior of elementary fields and their wavicles.

According to quantum field theory, there's another aspect of wavicles that's particle-like rather than wavelike: every wavicle has a definite rest mass. This isn't to say that every wavicle's rest mass is necessarily nonzero; some wavicles, most notably photons, have zero rest mass. My point is that the rest mass of a wavicle is meaningful and unambiguous.

We mustn't take this for granted; the world is full of things that lack an unambiguous rest mass. A rainbow, being an illusion, has no meaningful mass. The same is true of ideas, beliefs, and dreams, since these things have neither physics energy nor speed.

A cloud of steam is more concrete, but still, is it defined as the collection of its water droplets, or should the air inside the cloud be included, too? Either way, a steam cloud is continuously growing or shrinking, creating all sorts of ambiguities.

For an object to have a meaningful rest mass, it has to be distinguishable from everything else; it must be clear what is in it and what is not. It also has to be something whose speed can be clearly defined, since we need to know when it is stationary and to determine how its speed changes in response to a push. Rocks and balls certainly satisfy these criteria and have definite rest masses. It's natural to expect ordinary particles, such as grains of pollen and sand, to have rest mass, too, as these grains are like tiny balls.

By contrast, we would not naively expect waves to have a clearly defined rest mass. Familiar waves that slosh, spread out, and disappear are constantly undergoing changes that make it difficult to say where they start and end. Common sense would suggest that defining the rest mass for such a mercurial object would be a fool's errand; the result would be imprecise and subject to controversy.

For electrons and other elementary wavicles, however, there are no such ambiguities. An electron is an electron is an electron: it's a wavicle in the electron field, and so it operates as an unbreakable unit, making it a clearly defined object with a definite amount of energy. Under suitable conditions, its motion is easily assessed, too, making the measurement of its rest mass straightforward.

Does a medium have a rest mass? For an ordinary medium, such as an atmosphere, an ocean, a chunk of iron, or a rocky planet, the answer is yes. But this is not so for an amotional everywhere-medium. Its speed cannot be measured—it's amotional, after all—and we can't hope to push on it, either, since we can't even detect it directly. As for fields, even ordinary ones, they never have a rest mass; speed is for objects, not for their properties. (My car certainly has a speed, but its age, length, and resale value do not.)

Photons deserve special commentary. Since we can't ever make them stationary, we can't measure their rest masses in the usual way. Instead, we may appeal to a fundamental rule of Einstein's relativity, stated all the way back in Chapter 5: only objects with zero rest mass can travel at the cosmic speed limit. Since that's the speed of light waves of all frequencies, all photons have zero rest mass. The same logic applies to gluons and to gravitons, the quanta of gravitational waves.

Thus, for each elementary field of nature, there's a corresponding type of wavicle, with a definite rest mass. For the electron field, there are electrons, every one of which has a rest mass identical to that of all the others. For the electromagnetic field, there are photons, all with zero rest mass. The W field has W bosons, each of which has the same mass.[1] Macroscopic objects are different; every raindrop and dust speck has its own unique mass. There's something special about elementary wavicles—something that makes them more reliable than other objects. It's time to track that down.

Toward that end, we at last bring in the quantum formula, $E = f[h]$. Perhaps you'd forgotten about it?

When Einstein proposed that light is made from quanta, he took the quantum formula that Planck had written down in 1900 and suggested a profoundly new way to interpret it. Planck, who introduced it while trying to explain how hot objects glow, suspected that it had to do with how *atoms* work. Einstein's proposal was that this formula tells us how *light* works. That was bold and surprising. Physicists thought they already knew how light works...and it wasn't like this!

In hindsight, we may describe Einstein's proposal as follows. Take a simple light wave with a frequency *f*, such as a laser beam. That wave is made from wavicles—photons—each of which has exactly the same amount of energy. The energy *E* carried by each photon is given by the quantum formula: it's the frequency *f* times the conversion factor *[h]*.

What is *h*? In honor of Planck, who discovered it, *h* is called *Planck's constant*. It's a constant of nature as important as *c*. While *c* is a cosmic speed limit, *h* is a sort of particle-ness limit and more generally a *cosmic certainty limit*. (I think I'm the only one who calls it that. But I think it captures how *h* and *c* play analogous roles.)

The implications of *h* are so vast that entire bookcases in scientific, historical, and technological libraries are devoted to it. But for our story, the important part of the quantum formula is everything else: the relationship between *E* and *f*, between energy and frequency. For this reason, I will merely give a cursory overview of *h*.

To understand what I mean by a "particle-ness" limit, let's examine what we usually imagine a particle to be. Roughly speaking, we conceive of it as similar to a ball, only smaller. The smaller the ball, the sharper and narrower its path as it moves around—its *trajectory*—will be. An ideal particle, infinitely small, would have a razor-sharp trajectory, as shown at far left in Fig. 40.

But in a quantum universe, you can't make an object's trajectory arbitrarily sharp; there's a fundamental limit. Even though an electron is incredibly small, it's not a dot, and like any wave, its general tendency is to spread. Its trajectory must therefore be much fuzzier than its size would suggest, at least on average. The very process of measuring its position momentarily counters this spreading, and so, if we repeatedly make gentle, imprecise measurements of the electron's position, being careful not to try too hard, we can encourage it to keep a near-constant fuzziness, as in Fig. 40, left of center. This is as particle-like as we can ever make its trajectory over time.

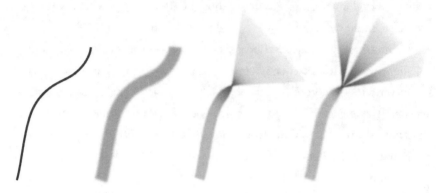

Figure 40: (Left to right) A particle of zero size has a sharp trajectory. But a wavicle's path, measured gently, is fuzzy. A collision with another particle may shrink an elementary "particle" to an extremely tiny size, but immediately thereafter, it will spread out. If a particle with a finite size, such as a proton, undergoes a similar collision that shrinks it too far, it will slosh internally and will be converted into multiple particles.

If a wavicle is always spread out, and can be more or less spread out at different times, what do we even mean when we say that the electron has a size? And how can we measure that size? Through a suitable experiment, we can try to shrink a wavicle down *for a moment*. If it is really infinitely small, we should be able to squeeze its trajectory down as far as our current technology allows us. But still, we can do it only for an instant; the more compact the wavicle's trajectory becomes at one moment, the fuzzier it becomes immediately thereafter, as in Fig. 40, right of center. In a similar sense, we could squeeze an ocean wave to make it briefly into a sharp spike of water, but it would promptly squirt out in all directions. There's no way to give any wave, even a wavicle, a sharp trajectory.

The same method can also reveal a proton's finite extent. Thanks to *h*, a proton, too, can spread out. But any attempt to make it significantly smaller than what we refer to as "the proton's size" will disturb its interior. The resulting internal disruption will quickly lead the proton to disintegrate into multiple particles, as at far right in Fig. 40. This difference between electrons and protons confirms that the latter has a measurable size, while the former (up to now) apparently does not.[2]

You might infer that the cosmic certainty limit means that certain aspects of electrons and other wavicles (such as their trajectories) can never be perfectly known. This notion is captured in what is known as Werner Heisenberg's quantum uncertainty principle. When the principle is applied to an individual wavicle, it implies that its position and its motion cannot simultaneously be known with high precision, and this, in turn, implies that it can't have a sharp trajectory.

But in this particular context, it's perhaps better to say that certain aspects of our universe, ones that our common sense would have thought knowable, simply do not exist. It's not just that we can't precisely measure an electron's path, which you would expect a particle to have. An electron is a wavicle, and wavicles can't have sharp, well-defined paths.[3]

Knowing now that electrons are wavicles rather than dots, we ought to reconsider what it means to say that "atoms are mostly empty space." Remember my analogy that an atom is as empty as a classroom with nothing in it but a grain of sand. It's not exactly true.

As a wavicle, an electron in an atom spreads out around the atom's nucleus. The electric pull from the nucleus wants to draw it closer, but its wavicle nature wants it to spread, and a balance is found between the two.

This might tempt us to view the region around an atomic nucleus as full, packed with electron wavicles. But this would give us the wrong intuition. The electrons themselves are nothing but vibrations in the electron field; they aren't solid, impenetrable objects. Even with the electrons there, the region outside the atomic nucleus still acts, for most purposes, as if it is mostly empty.

We're used to the idea that waves needn't be impenetrable. You and I easily cross rooms that may be swarming with microwaves from our cell phones and radio waves from local radio stations. Similarly, if you direct a neutron to fly across an atom, it will generally go straight through; the likelihood that it will interact with any of the electron wavicles on its way is extremely small.[4]

In short, even though electrons spread around their atomic nuclei and in a sense do occupy the whole atom, they don't really make it full. Instead, their presence is benign and makes the space only slightly less empty. My earlier statements about atoms—first, that X-rays, neutrinos, and empty space itself can go right through them (and they through empty space), and second, that atomic matter can be crushed down dramatically to make a neutron star—require no changes.

Because of this, merely replacing the dot-electron with the wavicle-electron doesn't explain why atoms are impenetrable to one another. We are still missing an important insight from quantum field theory, so we will have to return to this puzzle later.

Like the relativity formula, the quantum formula is astonishingly simple and universal. Einstein initially applied it to light, but today we understand that it applies to *all* quanta of *all* vibrations. That includes all wavicles of fields, both ordinary and cosmic, as well as the minimal vibration of a swinging pendulum, a ringing bell, a rippling string, and anything else we've talked about.

It's frequency alone that determines the energy of a quantum. You do not need to know what's vibrating or why. If a steel guitar string, an aluminum bell, an electromagnetic wave, and a quartz resonator are tuned to the same note—if they vibrate at the same frequency f—then the quanta of their vibrations have exactly the same energy f [h], even though the vibrating objects are otherwise different in every respect. Nothing but frequency matters: not shape, material, strength, mass, age, etc.

Einstein's interpretation of the quantum formula explains many phenomena involving light, visible and otherwise. Among the most important is this: *a bright (i.e., high-amplitude) beam of low-frequency light won't hurt you nearly as much as a dim beam of high-frequency light.* In other words, a large number of low-frequency photons can't do nearly as much damage as a small number of high-frequency photons. "Better to be struck by a thousand ping-pong balls than by one bullet," as one of my teachers explained.[5]

When the atoms in your body absorb light, each does so one photon at a time.[6] If the photons are weaklings, with low frequency and therefore low energy, each photon is harmless to the atom that absorbs it and to those nearby. A horde of photons may strike a horde of your atoms, but they won't injure you (unless the number of absorbed photons is so high that your skin becomes hot). A floodlight of visible photons poses no risk; radio wave photons are even less worrisome.

But a photon with higher frequency has more energy, by the quantum formula. Each photon of ultraviolet light has enough energy to rip apart an atom, damaging one of your cells. This makes a relatively small amount of ultraviolet light much more dangerous than a great deal of visible light. X-ray photons are so powerful that they can damage many cells at once, which is why, when you need to have an X-ray image made of one part of your body, the rest of it is often shielded with a lead blanket.

Einstein used the quantum formula to explicate the *photo-electric effect*— the observation that dim high-frequency light can kick electrons out of the atoms of a metal, but bright low-frequency light cannot. Because light is made from photons that are absorbed one by one and not in parts, the small amount of E found in each low-f photon is insufficient for it to expel an electron, whereas the high E in a high-f photon gives it the oomph to do the job. If an atom could absorb a fraction of a photon, or could easily absorb ten photons simultaneously, this would be very hard to explain. For this imaginative idea, Einstein won his lone Nobel Prize.

17.1 The Energy Within

Finally, we are ready to combine the relationships embodied in Einstein's quantum and relativity formulas. This will lead us to a third formula, along with fresh insights. But before we plunge ahead, let's take a step back and remember how we got here.

Mass is intransigence, and not the same as weight; so Newton taught us. Einstein then realized that mass comes in different varieties. Though some of them depend on one's perspective, an object's rest mass—the intransigence of an initially stationary object—does not. Its origin is the object's internal energy, as revealed in the relativity formula written as $m = E/[c^2]$. For a proton, this E really means the energy *inside* it, but for an electron, this is yet to become clear.

Now, we face the questions from many chapters ago posed by my skeptical student. Elementary wavicles supposedly are of zero size and have no inside in which to store anything. How can they have internal energy? Where does that energy come from, and where and how is it stored?

It struck me, as I was writing these paragraphs, that my own teachers never directly addressed these issues. The resolution to these puzzles showed up in some math one day, hidden in the middle of a long technical analysis, and nobody said anything about it. It's only in retrospect that I recognize it as an interesting apparent paradox.

Now, let's resolve it.

The energy that gives the electron its rest mass isn't stored inside like water in a jar. Instead, it's a part of the electron, so essential to the electron's very existence that it can't be removed even in principle. Rather than *internal*, a better word might be *inherent*.

To clarify this, let me first propose a crude visual image of what a stationary electron might "look" like. (You can't actually hope to see it—the very act of shining light on it will move and dramatically change it. But there's still something useful about having an image in mind.)

"How might we visualize an electron that's motionless?" I asked a class of non-science students. "When I shine a laser pointer across a room, the laser light is a traveling wave in the electromagnetic field. A single photon from that laser is a traveling wavicle. In a similar way, an electron moving across the room is a traveling wavicle in the electron field. So..."

"I see," said a student, her eyes lighting up. "A stationary electron is a *standing* wave?"

"That's right," I confirmed. "A standing wavicle in the electron field. Imagine the electron field, waving in place."

"So is it like a vibrating guitar string, then?" asked another student, frowning.

"Somewhat," I replied, glossing over the fact that the guitar string is a medium, whereas we have to use the field-centric perspective here. "But there are two important differences.

"First, we have to imagine drastically modifying the standing wave on a guitar string by sharply diminishing its amplitude, greatly broadening its wave crest, and increasing its frequency. The wave extends away from us, tailing off gradually and disappearing out of our view; we can no longer see its ends." I drew on the board what is shown at the bottom of Fig. 41: a wave with a long, low contour.

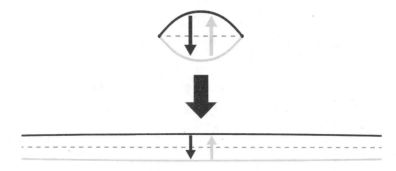

Figure 41: Toward visualizing a stationary electron: start with the standing wave shown at the top, and then stretch the wave horizontally, shrink it vertically, and increase its frequency to obtain the broad standing wave shown at the bottom.

"Second, instead of a wave on a string that extends in only one direction, our electron is a wave that extends away from us in all directions, a vibration of the electron field that reaches out to distances far larger than an atom.

"That gives us the best image I know of: a high-frequency low-amplitude extended standing wave stretching off in all directions."[7]

This image of a stationary electron isn't something I can draw. But it somewhat resembles a sound wave resonating in a closed room, except without the room—without any walls. (Similar waves will appear in Chapter 20, along with more explanation.) Whether or not you can visualize that, the lesson I want you to take away is this: *it's nothing like a stationary dot.*

"Notice how different this is from the image in the atomic cartoon," I remarked, "which encourages us to imagine a stationary electron as a little ball, sitting there and doing absolutely nothing. A standing wavicle is much more interesting than that. It's always *vibrating*. In fact, an electron is vibrating a million billion times every second.

"And here's the key point. Simply because it's vibrating, *it has energy!*"

The first student squinted at me. "Are you saying that the electron's vibration energy is what gives it rest mass?"

"Exactly!" I exclaimed.

Now the rest of the class was starting to look interested. After a few moments to let them think, I gave them another perspective.

"Let me say this a different way. Imagine the electron field in your vicinity isn't vibrating; its value is zero, and the space is empty. You decide you want to make an electron from scratch. To do that, you have to make the electron field vibrate, like a guitar player plucking a string. The amount of energy you need to add to the electron field, to get it to vibrate with precisely one stationary electron—one wavicle's worth of vibration, standing still in front of you—is exactly the electron's rest mass times $[c^2]$.

"Here's yet another way to look at it," I added. "An electron is, by nature, a vibration. It can't exist without constantly vibrating, any more than a human can exist without a heartbeat. All vibrating things have energy, even if they're not going anywhere. So an electron, even when stationary, has to have energy. This energy is intrinsic to it; take the energy away, and the electron ceases to exist. It's this intrinsic energy that gives it rest mass."

An electron is a vibration; vibrations always have energy; without the energy, there is no vibration and therefore no electron. It's the vibrational energy that leads to rest mass. I summed it up in a way I hoped the students wouldn't forget.

"In essence, *an electron's rest mass is its energy-of-being.*"

17.2 Resonance and Rest Mass

This is the long-awaited secret. Our bodies are constructed from vibrations—*literally*. These vibrations are quanta of waves, known as wavicles. And an elementary wavicle's rest mass arises from its energy-of-being…the energy required for it to exist in the first place.

This energy isn't like the chemical energy stored in a battery or the fuel in a car, which can be added or taken out. Nor can it flow in and out like water. It's a precondition for the electron to be.

You might reasonably complain that I ought to have explained this from the start. I've misled you for half a book by implying that an object's rest mass is always the energy that is found inside it. In my experience, telling the whole story too early makes it harder to understand. You have to first escape the atomic cartoon. If it's still dominating your thinking—if you're visualizing an electron as a dot—the possibility that the electron could have energy *inherently*, by the very nature of its vibratory existence, doesn't make any sense. Scientists themselves took decades to figure it out.

It's quite natural, from the perspective of someone looking at an electron from a distance, to misinterpret its hidden vibrational energy. It clearly isn't motion energy, because it's there even when the electron is stationary, and it clearly isn't external to the electron, since it travels with the electron wherever it goes. Based on common sense, one might guess that this energy must be internal to—stored inside—the electron. It takes imagination to realize that the electron could possess this energy in a completely different way, as only a wavicle can.

We've now learned where a wavicle's rest mass comes from, but we haven't learned how much it gets and why. Nor is it clear yet why all wavicles of a given type have the same rest mass.

Recall the main difference, as far as frequency, between traveling and standing waves: traveling waves can have a wide range of frequencies, but the simplest standing waves must vibrate at a vibrating object's resonant frequency. You were even able to check this directly if you performed the do-it-yourself experiments on waves that I suggested.

The quantum formula teaches us that a wavicle, whether traveling or standing, has energy E equal to its frequency f times a conversion factor, Planck's constant $[h]$. The higher the frequency, the greater the amount of energy the wavicle carries.

If a wavicle is traveling, its frequency can be anything—f can be as big as you want, and the quantum formula then tells you that E can be as big as you want. Particle accelerators make beams of swift, highly energetic electrons all the time.

But if, as required for a measurement of its rest mass, you make the wavicle stand? Well, then its frequency has to be the resonant frequency of its field, in analogy to the standing wave on a plucked guitar string.

Now, relativity and quantum physics come together to make something new. We have reached the central moment of this book.

The electron field has a resonant frequency, which is what the letter f will denote for the next few paragraphs. Therefore, it has standing waves that vibrate at that frequency. A standing wavicle of the electron field—a stationary electron—must therefore have vibrational energy $E = f[h]$, according to the quantum formula. That vibrational energy $f[h]$ is the wavicle's energy-of-being.

According to the relativity formula, a stationary object with energy E has rest mass equal to its energy divided by $[c^2]$. Since a stationary electron has energy-of-being $f[h]$, its rest mass is $f[h]$ divided by $[c^2]$.

And so the electron's rest mass m is equal to $f[h]/[c^2]$, where f is the electron field's resonant frequency.

This relation between m, the rest mass of an elementary "particle," and f, the resonant frequency of the corresponding field, isn't limited to electrons. It holds for all elementary wavicles of all elementary fields—quarks, neutrinos, W bosons, and all the others. For every field with a resonant frequency f, the rest mass m of its wavicles satisfies the same relation:

$$m = f\left[\frac{h}{c^2}\right],$$

where I have combined the two conversion factors into one.

Like Einstein's two formulas, this one is again a fundamental relation between two quantities, rest mass and frequency, with a conversion factor $[h/c^2]$ of importance only to those who need to use it explicitly. That said, the "=" sign means something more limited than in either the relativity or quantum formulas separately; we can't combine the two formulas willy-nilly. Mass is not equal to frequency in some general sense but only in a very specific one. *For an elementary wavicle, rest mass represents its energy-of-being—the energy required for the wavicle to exist—which in turn is set by the resonant frequency of its field.*

This explains, finally, why rest mass has everything to do with resonance. Fields that can resonate have wavicles with rest mass. For a species that so often puts music at the center of celebration, worship, and pleasure, this is deeply appealing. Like any musical instrument, the cosmos resonates with a pattern of frequencies, one that our formula translates directly into the pattern of the elementary wavicles' rest masses. These wavicles, the bricks of the material world, are the musical quanta—the quietest tones—of this instrument. The universe rings everywhere, in every thing.

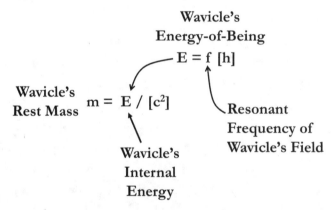

Figure 42: The origin of the relation between *m* and *f* via the combination of the relativity and quantum formulas.

These revelations allow us to resolve several mysteries. First and foremost, there is the question of how every single one of the myriad electrons scattered across this vast universe could have exactly the same rest mass. Well, now we know. It's for the same reason that a guitar string always produces the same note. Resonance.

Every time you pluck a guitar string, it vibrates with its resonant frequency. In the same way, the electron field's standing waves always vibrate with the field's resonant frequency. Since stationary electrons are standing wavicles, they always have the same vibrational frequency f. By the quantum formula and the relativity formula, that means they always have the same rest mass m. So long as the electron field's resonant frequency remains fixed, all electrons everywhere will always have the same, unchanging rest mass. The only way to change m would be to change f.

What's true for the electron is true for all the elementary wavicles with rest mass. Each field has its own unique resonant frequency, which then gives all its wavicles the same rest mass.

In fact, every electron's intrinsic properties, not only its rest mass but also its electric charge, its "spin," and its tendencies to interact with certain other fields, are the same as those of all other electrons. That's because each electron is exactly the same type of thing: a quantum of the electron field. More generally, the wavicles of any type are all identical to one another and inherit all their intrinsic properties from their field. This includes not only quarks and Higgs bosons but also wavicles without rest mass: every photon has the same intrinsic properties as every other photon, every gluon has the same intrinsic properties as every other gluon, and so forth.

This rigidity of form is why electrons and quarks and gluons, unlike rocks and stars and human beings, never get old. Age involves wear and tear, loss of integrity, damage. You can't damage an electron; it doesn't accumulate scuffs and scars from the buffeting that it may have experienced throughout its life of perhaps billions of years. It can't. It remains, always, a single quantum of the electron field, period.

Less obviously, the same is true for protons and neutrons. This is because they are formed so directly from the basic wavicles of the universe. We'll return to this point in the book's final chapters.

The importance of these facts for human life cannot be overstated. If electrons weren't identical, then each atom of oxygen would be different from every other, and our bodies would face a major quality-control problem with every breath. Badly damaged oxygen atoms might be useless or even dangerous to human health, and our bodies would have to filter them out. That might not sound so bad until you remember how many oxygen atoms we inhale every time we breathe! The fact that all oxygen atoms are chemically interchangeable makes it much easier to sustain living creatures. It also helps human engineers, who can create sheets of pure aluminum without having, say, to carefully remove older atoms that just aren't as strong as when they were new.

There is no garbage dump for wavicles—no repair shop, no retirement home, no hospital for recovery. The wavicles of nature are infinitely recyclable. How different a world it would be otherwise.

In fact, the exact identity of electrons is crucial for all of atomic physics and chemistry. That's because of something called the Pauli Exclusion Principle, which states that if two fermionic wavicles are strictly identical, they cannot do the same thing at the same time. (For now, just as in most chemistry classes, we'll view this rule as ad hoc; we'll return to its origins later.) This means that within atoms with many electrons, no two electrons can behave identically.

Because of this, assembling an atom's many electrons is analogous to assigning people to chairs in a sloped auditorium. First, you fill the lowest seats; then you put the next people in the row above that, which takes more energy since they have to climb stairs; and then you fill the row above that one, which takes even more energy. If only people were willing to sit in each others' laps! Then you could put all of them in the lowest row, with a significant energy saving. If electrons were bosonic wavicles, they'd happily pile on top of one another. But the fermionic electrons of our universe won't do it.

These rows and seats are called *shells* and *orbitals* by scientists. Because each type of atom has a different number of electrons, it fills its shells and orbitals in a unique way. Most atoms leave their highest shell partially empty, and their chemical properties depend crucially on how many orbitals in that shell remain open. Carbon and silicon, with a half-filled highest shell, have especially rich chemistry; fluorine and chlorine, with only one orbital remaining in their highest shell, easily form acids; and so on. Indeed, the pattern of filled shells and orbitals determines the shape of the periodic table of the elements. If electrons weren't exactly identical, or if they were bosonic instead of fermionic wavicles, none of this would happen; they would all congregate close to their nuclei, making atoms both smaller and far less chemically diverse.

A last insight from our new formula concerns the relation between mass and size. Among the known wavicles, the top quark's rest mass is the largest, about 340,000 times greater than an electron's. The electron's rest mass is at least a million times greater than those of the neutrinos, which have the smallest nonzero rest masses. Perhaps it seems strange that objects of such wildly different rest mass could all have the same size, or no size at all.

Common sense implies that if you want to increase the mass of an ordinary object a hundred times, you'll have to make it many times larger, too. This intuition is fine for rocks, planets, and even black holes. But in the subatomic world of wavicles, it doesn't work. An elementary wavicle may be arbitrarily small, in the sense of Fig. 40 (p. 238), and yet have plenty of rest mass. Instead, a wavicle's rest mass reflects its field's resonant frequency. The enormous range of wavicle rest masses is matched by an equal range of frequencies, extending over at least forty octaves, for the elementary fields.

In the subatomic world, then, mass requires a completely different intuition from what common sense provides. If you're an ordinary object and want more mass, fatten up. But if you're an "elementary particle" and you want more mass, sing higher.

18

Einstein's Haiku

As I launched the Fields portion of this book, I posed a set of paradoxes. We can summarize them this way: How can waves, which require a medium, and particles, which prefer empty space, both travel freely across the cosmos? If space is as empty as it seems, and as empty as the coasting law and Newton's view of orbits imply, why can light waves cross it? If instead it contains a medium for light, why doesn't that medium ruin Galileo's principle and the coasting law, producing drag on objects that pass through it and giving us a substance relative to which we can measure our motion?

Well, now we have our answer. Sort of.

At first glance, the answer looks simple enough. Everything is waves; more specifically, every thing is made of wavicles. Nothing is made of particles, in the English sense of the word. Space is not, in fact, empty—at least, not in the usual English sense in which *empty* means nothingness. It's no longer clear that we should ask whether empty space *contains* a medium; instead, it *is* a medium. At a minimum, it is the medium for the gravitational field. That field is just one of many elementary fields. It's not clear whether the other fields have media; perhaps they do not. If they do, perhaps those media are somehow amalgamated with empty space. It is also possible, at a maximum, that empty space is the medium for *all* the elementary fields. This would require empty space to be suitably enhanced by imperceptible features, such as Kaluza and Klein's extra dimensions or some other internal structure.

Whatever the details, this answer comes at a high price for common sense. It requires empty space to be amotional so as to preserve Galileo's relativity and the coasting law, but this makes the empty space around and inside us undetectable, as though it is both there and not there. It also requires waves to be made of identical wavicles, discrete building blocks from which ordinary objects may be constructed, but this implies that these waves have particulate behavior and other miraculous capabilities beyond easy reach of our imaginations.

To preserve Galileo's principle, the waves and wavicles of the elementary fields obey peculiar rules that would be logically impossible for any ordinary objects, waves, or media. If these wavicles travel at the cosmic speed limit, they must do so from everyone's perspective; as in a nightmare, they cannot be outrun. Even if they travel more slowly, they must always respect the cosmic speed limit, equally from everyone's perspective. It is thought-provoking that all these waves obey the same speed limit and that their fields are found everywhere in the cosmos. It hints that the fields may all be melded, along with empty space, into some kind of unified edifice, or framework, or ... well, universe.

This book is far from over. But everything up to now was aiming at this past chapter, and the rest will be built upon its foundation. So let's take a moment to contemplate all the distance that's been traveled, not just in this book but by our species, along one path from relativity and motion to mass and along a second from waves and fields to quanta. At the crossroads stands the cosmos.

Einstein's Haiku
E equals f h,
And E equals m c squared;
From these seeds, the world.

Field	Wavicles	Frequency Ratio Mass Ratio	Long-Lived (> 1 second)
Top quark field	Top quark/anti-quark	340,000	No
Higgs field	**Higgs boson**	240,000	No
Z field	**Z boson**	180,000	No
W field	**W+/ W- bosons**	160,000	No
Bottom quark field	Bottom quark/anti-quark	8,200	No
Tau field	Tau/anti-tau	3,500	No
Charm quark field	Charm quark/anti-quark	2,500	No
Muon field	Muon/anti-muon	210	No
Strange quark field	Strange quark/anti-quark	170	No
Down quark field	Down quark/anti-quark	10	Yes
Up quark field	Up quark/anti-quark	4	Yes
Electron field	Electron/positron	1	Yes
Neutrino fields (3)	Neutrinos (3)	less than .0000002	Yes
Gluon field	**Gluon**	0	Yes
Electromagnetic field	**Photon**	0	Yes
Gravitational field	**Graviton**	0	Yes

Table 6: The known fields, along with their frequencies divided by the electron field's frequency (and thus their wavicles' rest masses relative to the electron's), and noting whether their wavicles exist for longer than one second. Bosonic fields are in boldface. Numbers are approximate. The three neutrino rest masses are known to be small; at least two are known to be nonzero, with none precisely measured yet. In the last three rows, the "0" really means "too small to measure, and believed to be zero."

HIGGS

Knowing now what rest mass is for "particles," we are finally in a position to understand how the Higgs field makes its mark upon the universe. But first, let's recall why this hitherto secretive field merits such an extended discussion in the first place.

Early in this book, I dismissed the tendency of certain journalists and science writers to refer to the Higgs boson as the "God Particle." It oversells the particle and annoys the physicists (including Peter Higgs himself[1]). But if physicists dislike it, where did the name "God Particle" come from?

It's as you'd expect in our materialist society. Advertising. Marketing. In modern parlance, it's clickbait.

In 1993, Leon Lederman, who'd won a Nobel Prize for leading the team that discovered the bottom quark, cowrote a book with science writer Dick Teresi about the ongoing search for the Higgs boson. To sell the book, either the authors or their publisher invented the title *The God Particle: If the Universe Is the Answer, What Is the Question?*

To my mind, this is truly an impressive piece of marketing, one that future advertisers ought to study in school, because it manages so effectively to insult both science and religion while seeming to promote and unite them. Though the Higgs field plays a crucial role in the universe, the Higgs boson does not, and it hardly deserves this grandiose name. Meanwhile, the idea that divinity could be captured in a wavicle, one that exists for a ridiculously tiny fraction of a second, is beyond absurd.

When I complained about this to a friend, he quoted "Let there be Light" and suggested the photon as a better candidate for the God Particle. In a competition, other natural nominees would be neutrinos, which help heavier atoms form in stars, and the elementary wavicles found in atoms. But the Higgs boson? All that particle has going for it is a top-notch public relations team.

Now, should we call the Higgs field the "God Field"? I don't recommend it. But at least the field, unlike the particle, is truly of cosmic importance!

In the coming chapters, we'll see how the Higgs field generates wavicles' rest masses and explore the puzzles raised by its existence and behavior. But let's anticipate what has to happen. If the electron is to get a rest mass, then we know from Chapter 17's formula that the electron field must get a resonant frequency; you can't have one without the other. So the Higgs field's nonzero average value must somehow alter the resonant frequency of the electron field. From the discussion in Chapter 10.2 about changing the resonant frequencies of vibrating strings, we might then suspect that the Higgs field has an impact on the electron field's environment.

Somehow it has to accomplish this without ruining Galileo's principle. That's not something most fields could do.

= 19 =

A Field Like No Other

Some fields "point," and some don't. The wind field and the magnetization field are examples of pointing fields; they have both a strength and a direction, and were accordingly depicted with arrows in Figs. 32 (p. 175) and 34 (p. 185). Check your favorite weather website: the wind in your town may be stated as 20 miles per hour from the south or as 10 meters per second from the northeast, but it always comes as an amount *and* a direction of origin.

Pressure, by contrast, is a nonpointing field. It's given purely as an amount: the air pressure might be stated as 30 inches of mercury or perhaps 980 millibarns. The same is true of a density field. The mass density of olive oil, 0.92 grams per cubic centimeter, needs no orientation.

Among known elementary fields, the Higgs field is the only one that doesn't point.[1] This turns out to be decisive in allowing it to generate mass without messing up anything else.

Forces also point; if you throw a ball or pull open a drawer, you exert a force in a particular direction to do it. For a field to create a force, it has to specify how the latter should point. This is easy for the wind field: if a steady wind blows from the west, any trees that it knocks down will fall to the east. Similarly, our local gravitational field, which points down, creates a pull toward the ground. But even with a nonzero value, a field that doesn't point can't create a force on objects; a force needs a direction, but a nonpointing field has none to give it. For example, constant, uniform pressure pushes on objects but does so equally from all sides; consequently, it produces no net force and has no net effect on their motion.

If pressure varies from place to place, however, that *variation* itself points from high pressure to low, and now there can be a force. For instance, as in Fig. 43, if the pressure on a wall is greater on its left side, the wall will be pushed to the right. The larger the pressure difference, the greater the force will be. Pressure differences induce many familiar forces, such as those that inflate a balloon or hurt your ears when you ride a fast elevator.

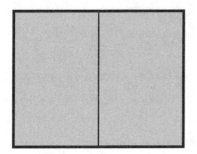 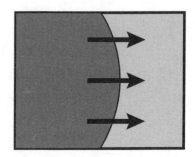

Figure 43: (Left) With the same pressure on both sides of a wall, there's no force on the wall. (Right) If the pressure is stronger on the left, then the wall experiences a force (black arrows) that pushes it to the right.

As we'll see shortly, the Higgs field, too, can create a force when it varies from one place to another. But a constant, uniform Higgs field, such as you and I have been surrounded by all our lives, never creates a force. That's why humans across history never noticed it, just as they long failed to notice air pressure. Nonpointing elementary fields are unique in their ability to remain hidden.

Pointing fields can't hide so easily. If the electromagnetic field were nonzero and uniform across the universe, many objects would feel an electric or magnetic force no matter where they went. They would never coast. Such a universal breakdown of the coasting law could hardly go unnoticed. (Other more subtle breakdowns of Galileo's relativity principle would also occur.[2])

The Higgs field, by contrast, can have a nonzero average value without obscuring the principle of relativity. Its appearance is perspective-independent;

all observers, including those in isolated bubbles moving at any speed, see it as having the same nonzero value everywhere. This means that it can generate effects that are perspective-independent, such as shifting a field's resonant frequency and its wavicles' rest mass. Only a nonpointing cosmic field with an amotional everywhere-medium, or no medium at all, could achieve this. Any other type of field with a nonzero average value would ruin Galileo's principle from the very beginning.[3]

These differences between pointing and nonpointing fields become less significant when they vary across space. All bosonic fields, whether they point or not, can potentially act as intermediaries and generate forces between separated objects.

As an example, consider Earth's pull on the Moon. From the medium-centric viewpoint, gravity arises when empty space is warped. Though an object moving in flat empty space will coast in a straight line, an object traversing warped space will travel on a curved path. The Earth warps the space around it, more so close by than farther away, and an object that comes near and encounters that reshaped space will turn toward the Earth. We interpret this deviation from coasting as due to a gravitational force. The force weakens at greater distances, via the inverse square law, simply because Earth's leverage on space weakens there, too.

From the field-centric perspective on the same thing, we'd say that the Earth causes the gravitational field nearby to be nonzero. The field's value is large near the Earth's surface and dies off farther away. When an object approaches, it encounters and interacts with that nonzero gravitational field, causing its path to bend earthward.

The field-centric perspective explains many other forces between objects, such as the electric repulsion between electrons. Because the electromagnetic and electron fields interact, an electron both affects and is affected by the electric field. Around an electron, the electric field's value is nonzero, large near the electron and small farther away (as in Fig. 44 at bottom left). Meanwhile, an electron's path will change if it passes through a region where the electric field is nonzero. As two electrons draw close, each interacts with the electric field due to the other, altering its motion. We interpret the resulting

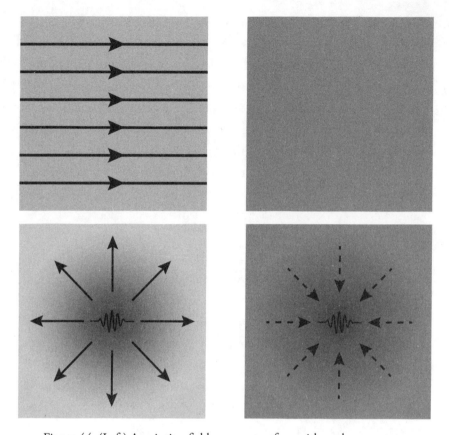

Figure 44: (Left) A pointing field can create a force either when constant (top) or when variable, as around a wavicle (bottom). (Right) A nonpointing field cannot create a force when constant (top) but can do so when varying around a wavicle (bottom). Arrows indicate the direction of the force; on the left, they also indicate the direction of the pointing field, and on the right, they indicate how the nonpointing field varies.

changes in the electrons' paths as due to an electric force that pushes them apart.

In a similar way, electrons can attract each other via the Higgs force. Though a uniform Higgs field won't create a force, the Higgs field around an electron isn't uniform (Fig. 44 at bottom right). Because the Higgs field and the electron field interact, an electron distorts the nonzero Higgs field around it, especially close by, and it reacts to any variation in the Higgs field in its vicinity, including one caused by a second nearby electron. The result is that the two electrons are pulled together.

Admittedly, this is academic; the attractive Higgs force between electrons is totally dwarfed by their electric repulsion, and no one is likely ever to observe it. In fact, the Higgs force is tiny to the point of irrelevance for all atoms and for all their subatomic components; as we'll soon see, this is related to the small rest masses of the electron and the up and down quarks. However, for top quarks and for W and Z bosons, whose rest masses are large (see Table 6 on p. 253), the Higgs force between them can be as powerful as any other force they experience. Most likely, the Higgs force will first be observed experimentally through the attraction that it induces between a top quark and a top anti-quark.[4]

The Higgs Field in Action

It's a blisteringly hot day in summer, and a stranger offers you an apparently derelict guitar. It has long, dangling strings that flop like cooked spaghetti when you try to pluck them. You're not inclined to accept it, but the stranger claims that the strings are imbued with magic. Put the guitar in a freezer, says the stranger, and when the strings get cold enough, they'll suddenly stiffen up. After that, they'll resonate, and the instrument will make music.

Once upon a time, there came into being a universe. Searingly hot, it swarmed with wavicles. Among its fields was a Higgs field, gyrating wildly. But as this universe expanded and cooled, the average value of the Higgs field suddenly became nonzero. When this happened, many previously floppy fields became stiff; they acquired resonant frequencies, and their wavicles gained energy-of-being and rest mass. That's how the universe was transformed, through the influence of the Higgs field, into the quantum musical instrument it is today.[1]

"Are you implying that the Higgs field is a sort of *stiffening agent?*"

"Yes, that's right," I agreed. "That's a good way to think about it."

"So it's kind of like cornstarch, then?"

I hesitated. "Well, not exactly. Cornstarch is a substance that stiffens other substances. But the Higgs field isn't a substance. And it isn't stiffening substances—it's stiffening other fields. It makes those fields vibrate differently, and that has something to do with how particles get their mass."

My friend gave me a look of bewilderment. This was my fault entirely; my last sentence had been completely incomprehensible, since I hadn't told him anything about wavicles or about how rest mass is vibrational energy-of-being. But I promised him that someday I'd give him a clear, thorough explanation of the whole thing. And here we are.

The basic overview is this. Most known elementary fields started out floppy—without stiffness. By this I mean that they had no resonance frequency and no standing waves. But nowadays they are stiffened by the Higgs field; instead of flopping about lazily, they ring vibrantly. Their wavicles then obtain rest mass through our formula from Chapter 17.2 relating resonant frequency to rest mass.

The stiffening agents we encounter in ordinary life don't work quite the way the Higgs field does. That's unfortunate; if there were a simple analogue that everyone could immediately grasp, no one would have found it necessary to invent the Higgs phib. Lacking an obvious shortcut, I will proceed in steps, each step adding another layer of insight into how the Higgs field does its job.

I'll start in Chapter 20.1 with a loose yet instructive analogy that relates the stiffening of fields by the Higgs field to the effects of gravity on a pendulum. This example may already satisfy many readers. Better but more elaborate analogies follow in Chapter 20.2, though these are optional in that the rest of the book does not depend on them. Readers who find them too detailed or otherwise unnecessary can safely skip them or give them just a brief glance.

20.1 A First Analogy

The terms *stiff*, *floppy*, and *stiffening agent* aren't sharply defined either in English or in physics dialect. The easiest way to see what I mean by them here is through examples, so let's head straight in.

Our first example involves gravity stiffening the position of a ball at the end of a pendulum. It's a rough analogy in some ways, but in others, as we'll see, it's surprisingly good.

If you put a ball at the end of a chain out in deep space, where the gravitational field is essentially zero, the ball will float aimlessly. If you give the ball a little push, its position may slowly drift, but it won't vibrate back and forth.

However, as in Fig. 45, if you put it in a nonzero gravitational field, everything changes. Now it will hang downward (i.e., wherever the gravitational field points) and, if disturbed, will swing. This swinging indicates that the ball's position has become stiff.

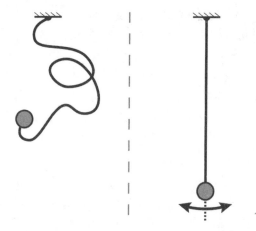

Figure 45: (Left) Without gravity, a pendulum won't swing. (Right) The gravitational field acts as a stiffening agent on the ball's position.

When hanging straight down and stationary, the ball is said to be *in equilibrium*—stable, balanced, and with no reason to go anywhere. If the ball is displaced to the right of that spot, it will swing back to the left; if displaced to the left, it will swing back to the right. This tendency for the ball's position to return toward the point of equilibrium is indicative of what is known as a restoring force or, more generally, a *restoring effect* (as it does not always involve a force). This is the effect that's responsible for the stiffening of the ball's position; it prevents the ball from drifting and causes it to vibrate if it is shifted away from equilibrium.

In this example, the gravitational field acts as a stiffening agent. Without it, there'd be no restoring effect, and the resonant frequency of the pendulum would be zero. The larger the gravitational field's value, the stronger the restoring effect and the higher the pendulum's resonant frequency. Keep this pattern in mind.[2]

Notice that the gravitational field has not stiffened the ball itself; the ball hasn't changed. Instead, gravity has stiffened the ball's *position*, which we may view as one of the ball's properties. This is parallel to the action of the Higgs field, which stiffens fields rather than their media (or other material objects) and in this sense affects properties, not substances.

More specifically, much as gravity creates a restoring force on the ball's position and pulls the displaced pendulum back to equilibrium, *the Higgs field creates a restoring effect on other elementary fields* that draws these fields' values back toward zero.[3]

Gravity's restoring effect causes a displaced pendulum to swing resonantly. Analogously, the Higgs field's restoring effect, countering any nonzero value for the other fields and driving their values back toward zero, makes standing waves in those fields possible. The resulting waves are precisely of the sort we imagined for a stationary electron back in Chapter 17.1. (See Fig. 41 on p. 243 for my sketch of an analogous wave on a string. The examples in Chapter 20.2 will clarify why the waves take this unusual form.)

If the gravitational field's value were zero, the ball's position would be free to drift and would not vibrate. Similarly, if the Higgs field's average value were zero, most fields would lack the stiffness needed for standing waves. Disturbing such "floppy" fields would create only traveling waves.

Conversely, a larger value for the gravitational field gives the pendulum a higher frequency. For the same reason, the larger the Higgs field's value, the more powerful its restoring effect on other fields and the higher their standing waves' frequencies.

At a surface level, this analogy is a good one. First, it shows that a restoring effect can enable a new mode of vibration. Second, it illustrates how a restoring effect can arise from the average value of a field playing the role of a stiffening agent. Reflecting this, the math formulas governing the Higgs field's activities bear a striking similarity to the formulas for gravity's influence on a pendulum.

Still, the resemblance may be only skin deep. Just as birds, bats, and bees can all fly but display significant differences when we look more closely, there's no general reason to expect that parallel phenomena must have exactly parallel causes. When it comes to the Higgs field, there may be nothing analogous to the pendulum's details, such as its chain, its supports, or even the ball.

To emphasize this point, I'll now consider another example that shows how gravity can stiffen a ball's position in a completely different way. Imagine a ball rolling in a bowl, as in Fig. 46. Without gravity, the ball can just sit anywhere on (or even off) the bowl. But in a nonzero gravitational field, all the symptoms of stiffness appear. Gravity draws the ball back to the bowl's center, the point of equilibrium, via a restoring effect analogous to the one seen in a pendulum. As before, stronger gravity produces a more powerful restoring effect and increases the vibration's frequency.

Superficially, the ball's back-and-forth motion in the bowl appears much the same as for a pendulum. Even the math formulas describing the two systems are similar (and become indistinguishable when the vibration's amplitude is very small). Only if we dig deeper will we notice the differences.

Although we've been focusing on stiffening agents, it's important to understand that they aren't always required. Certain types of fields can be intrinsically stiff, and their wavicles can have rest mass, without need of a stiffening agent. This is also true of properties of ordinary objects, as I'll exhibit in the next example.

At the bottom of Fig. 46, I've depicted a ball suspended between two horizontal springs. This ball is also subject to a restoring effect: its equilibrium point is at center, and if it moves to either side, the springs will draw it back. Here, in contrast to the previous examples, gravity plays no role. This ball's position is *intrinsically stiff*; it vibrates even in the absence of a stiffening agent.[4]

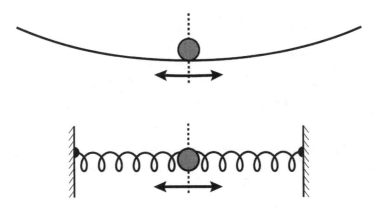

Figure 46: (Top) If a ball is placed in a bowl in the presence of gravity,
its position will be stiff. (Bottom) If the ball is held in place by two springs,
its position will be stiff even in the absence of gravity.

Let's take a step back now and consider the sorts of problems that scientists face in this business. I've drawn nice pictures to show you how these examples work, but suppose instead that the chain, bowl, and springs were invisible and that we could observe only the ball's motion and the value of the gravitational field. Could we still hope to understand, predict, and interpret the behavior of the ball? Yes, if we carried out a methodical, three-step scientific investigation.

First, we would ask whether the ball's position is stiff: Is it subject to a restoring effect? To find out, we would simply tap the ball and watch what it does. If it vibrates, we would know the ball's position has been stiffened by a restoring effect whose strength is reflected in the vibration's frequency. If it merely drifts, then there is no restoring effect and no stiffness.

Second, assuming we found that the ball's position is stiff, we would next ask whether it acquired its stiffness through a stiffening agent. By altering the gravitational field, we could investigate gravity's role. For instance, we could take the ball to a mountaintop, where gravity is a bit weaker, or even into deep space, where the gravitational field is negligible. If the vibration slows in weaker gravity and ceases altogether when gravity is absent, we would then know that the gravitational field is acting as a stiffening agent. Otherwise we would infer that gravity is not involved, leaving open the possibility that the ball's position is intrinsically stiff, as for the ball between springs.

Our final task, if we found that the gravitational field is the relevant stiffening agent, would be to determine whether the ball hangs from a pendulum, rolls in a bowl, or sits in some other complex setting. This would demand a much more elaborate set of experiments. We might begin our search for clues by studying, for instance, the influence of dissipation on the vibration's amplitude or the impact of a very large amplitude on the vibration's frequency.

For the elementary fields of the universe, our problems and questions are parallel to these, so we can proceed in a corresponding set of three steps. First, is a particular field stiff or not? That's easy to tell: if a field has wavicles with nonzero rest mass, then it has standing waves and must be stiff. Everything we need to know is already in Table 6 (p. 253); the electron field and quark fields are stiff, the electromagnetic field is floppy, and so on.

Our second step would be to ask whether the field is intrinsically stiff[5] or whether it has a stiffening agent.[6] This is a more challenging question than for the vibrating ball, because there is no practical way for us to alter the Higgs field's value to see how such a change would affect the field's standing waves.

The insights that led scientists to propose the Higgs field came slowly, from the 1950s and 1960s into the 1970s. Experimental studies of the weak nuclear force (for which the W and Z fields serve as the intermediaries) were critical, as were theoretical arguments that clarified the experimental results. These gradually made it evident that none of the elementary fields then known could be intrinsically stiff.[7] By the late 1970s, it was understood that all the stiff fields in Table 6, excluding the Higgs field itself, require a stiffening agent.

It is remarkable, and not at all obvious, that one and only one Higgs field can potentially serve as the stiffening agent for all of the other stiff fields in Table 6. (In many imaginable universes, the math for such a simple scenario wouldn't work.) This simplest of possibilities is called the *Standard Model*, or sometimes the *Minimal Standard Model* (MSM). Nature may be more elaborate than this, as I'll describe in future chapters. As of 2023, though, all data from the LHC support this minimalist option.

Finally, the third step in our investigations would involve diagnosing the underlying cause of the field's stiffening. This would require experiments that lie well beyond our present capabilities. We have seen that there are many ways to stiffen a ball's position; there are also many imaginable ways to stiffen fields. Much as simple experiments on a ball can confirm gravity's role but cannot distinguish a bowl from a pendulum, current particle physics experiments can do little except to confirm that the Higgs field acts as a stiffening agent. They are far from helping us learn how or why, assuming that such questions even have meaningful answers. To gain a more complete picture of what lies behind the Higgs field and its consequences, we may need to understand the origins of the universe's fields and the nature of empty space. Such insights do not seem close at hand.

20.2 Two Closer Analogies

The biggest limitation of my analogies so far is the absence of waves. Since wavicles are central in the Higgs field's story, let's now look at two examples in which waves appear directly. To reiterate, these examples play no role in the rest of the book, so a full appreciation of how they work is not essential. I hope many readers will find them useful, but you may also safely skip or skim this short section without any impact on later chapters.

First, we will replace the ball of the previous examples with a string, a one-dimensional ordinary medium. Then we will replace the string with a fully three-dimensional ordinary medium. The point of both these examples is to show why a stiffening agent's restoring effect leads to standing waves.

Any string can have traveling waves. A string with pinned ends can also have standing waves, including the simplest such wave with just a single crest. These two types of waves, which appeared in Figs. 23 (p. 143) and

25 (p. 145), are reproduced from a different visual perspective in Fig. 47, left and center. In both cases, the position of the string in the absence of any wave is shown as a dashed line; the waves involve vibration centered around this line.

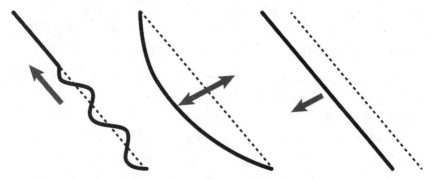

Figure 47: (Left) A traveling wave on a string; it may have any frequency. (Center) A standing wave on a string with pinned ends, like that on a guitar string; it vibrates with the string's resonant frequency. (Right) If a string with free ends (or an infinite string) is lightly pushed, it will drift away. In each case, dashed lines indicate the string's initial position.

A guitar string's pinned ends play a crucial role for its standing waves. Not only do they prevent the string from drifting away from the guitar, but they also help create a restoring effect, without which there would be no standing wave. If we loosen the string, or if we make it longer, the efficacy of the tied ends is reduced; the power of their restoring effect decreases, and so the frequency of the standing wave does, too.

If the ends are untied completely and allowed to float free, or if the string becomes infinitely long, the restoring effect disappears altogether. Then the standing wave's frequency drops to zero—that is, there's no standing wave at all. In that case, plucking the string causes only traveling waves. Correspondingly, a gentle nudge to the entire string will cause it to drift slowly away from its original location, as shown at the right of Fig. 47. There's no restoring effect to reverse its course.

However, we can change this by adding a new restoring effect that pulls the string back toward the dashed line. Then the string will exhibit a novel type of standing wave. We will do this by turning the string into an extended pendulum. Just as we might attach a ball to the bottom of a hanging chain, we will attach our entire string to the bottom of a curtain, similarly hung from the ceiling.

Even with this encumbrance, the string will still have traveling waves, as at the left of Fig. 48. If its ends are attached, it will still have a familiar standing wave, as in the center of Fig. 48. Qualitatively, these waves are much as they were without the curtain.[8]

But now imagine making the string infinitely long, or unpinning its ends. If we nudge the string as a whole, it will no longer drift away from the dashed line, as it did at the right of Fig. 47. Instead, thanks to the gravitational field and the curtain, the string will experience a restoring effect that pulls it back toward the dashed line, which now serves as a line of equilibrium. This is much the same as a pendulum bob pulled back to its equilibrium point. As shown at the right of Fig. 48, the entire string will swing back and forth as a unit, vibrating around the dashed line like a swinging pendulum.

Here the string is engaged in the most extreme form of standing wave, a uniform vibration across its full length. The wave is all crest and then all trough. It lacks even the familiar tailing off at the ends seen in a guitar

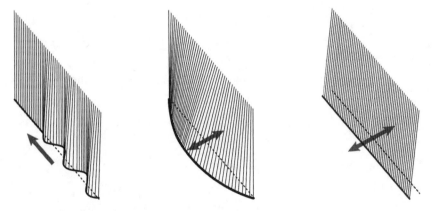

Figure 48: As in Fig. 47, but with a curtain attached to the string. (Left) Traveling waves are changed only in detail. (Center) The standing wave with fixed ends is changed only in detail. (Right) A string without pinned ends, or an infinite string, no longer drifts; as long as there is gravity, it now vibrates as a whole.

string's standing wave. Among all the waves seen in this book, it resembles most of all the standing wave I asked you to imagine for a stationary electron (Fig. 41 on p. 243).

This is no accident. Although I didn't say so at the time, the electron can vibrate this way only because it is subject to a restoring effect. That effect is generated by the Higgs field. It acts as the electron field's stiffening agent in much the same way as gravity acts as the stiffening agent for the wave at the right of Fig. 48. The analogy is close; even the math matches.

In this standing wave, in contrast to those on human string instruments, the length of the string and the status of its ends are irrelevant. The wave's frequency—the string's resonant frequency—is instead controlled by an aspect of its environment, namely, the value of the uniform gravitational field around it. The larger that value, the stronger the restoring effect and the greater the standing wave's frequency. Conversely, if the gravitational field's value fell to zero, the restoring effect would disappear and the string would drift instead of vibrate.

Back in Chapter 10.2, I mentioned that the universe has a trick up its sleeve, one not seen in human musical instruments, that allows an infinitely or immensely long object to have standing waves vibrating with a high resonant frequency. We have here an example of that trick. An environmentally generated restoring effect is the key ingredient. For the string, gravity produces the environmental effect; for the elementary fields, it's the Higgs field that does so.

Even if the curtain were invisible to us, we could apply our three-step scientific investigation. The existence of the string's standing wave would prove that its position has been stiffened. By studying how the standing wave responds to gravity, we would infer that the gravitational field is serving as a stiffening agent. More detailed studies would be needed to discover that there's a curtain involved in the process, too.

This analogy is closer to the reality of the Higgs field than those of the last section, both in concepts and in math. Still, the string exists only along a line, while the fields stiffened by the Higgs field are found everywhere in the universe, across all three dimensions of space. To bring us even nearer to the Higgs field, I'll conclude with an example in which a field of a three-dimensional medium is stiffened by the nonzero value of the electric field.

The process we'll explore parallels the behavior of a compass in a magnetic field. Just as Earth's gravitational field points downward and stiffens the position of a bob on a pendulum, the Earth's magnetic field points northward and stiffens the orientation of a compass needle (Fig. 49). When the needle aligns with the magnetic field, it is in equilibrium; otherwise a restoring force brings it back toward alignment. If the magnetic field's value were zero, there'd be no restoring effect, and the needle would be equally content to point in any direction.

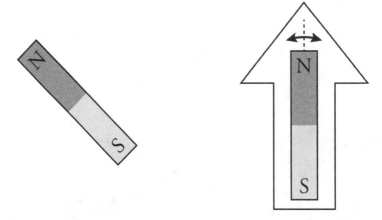

Figure 49: (Left) If the magnetic field nearby is zero, a magnetized needle's orientation angle is arbitrary; it can point in any direction. (Right) In the presence of a nonzero uniform magnetic field (large arrow), the needle aligns itself with the magnetic field, and its orientation angle is stiff.

In a similar way, there are needle-shaped molecules that naturally align with a nonzero *electric* field. Such molecules are found in certain types of *liquid crystals*, relatives of those used in liquid crystal display (LCD) screens.

As is typical of an ordinary medium, a liquid crystal has numerous fields. Among its properties that we can view as fields are its mass density, pressure, and flow. But the property we will focus on is the *orientation field* of the molecules, somewhat analogous to the magnetization field of a magnet, as in

Fig. 32 of Chapter 13. The molecules' pointing direction can vary across the material, which is what makes their orientation a field.

Although these molecules form a liquid and can move around and past one another, they often tend, because of their shape, to align with each other. At sufficiently cold temperatures, the molecules will become parallel, as at the top left of Fig. 50, making the orientation field's average value uniform and constant. But just as for a compass needle in a zero magnetic field, the orientation field could happily point anywhere. It has no equilibrium direction, and there's no restoring effect acting on it. If we have a vial of this liquid crystal in which the molecules are aligned toward the southeast, and we gently rotate the vial by 90 degrees, the molecules will all rotate together, leaving them aligned to the northeast. This transformation is depicted on the left of Fig. 50.

Now, let's make the electric field nonzero and pointing eastward, all around and through the liquid crystal. The interaction between the electric

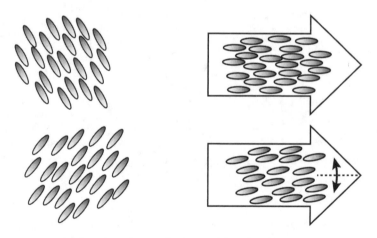

Figure 50: (Left) A liquid crystal's elongated molecules can move around but tend to align. The orientation of their alignment is a floppy field; two equivalent orientations are shown. (Top right) In a nonzero uniform electric field (large arrow), the molecules will align with the electric field. (Bottom right) The orientation field is now stiff; if the molecules are uniformly tilted away from the electric field, a restoring effect draws them back into alignment and a standing wave results (black arrows).

field and the molecules will cause the latter to rotate so that they align with the former, as in the top right of Fig. 50. The orientation field is now stiffened; its equilibrium pointing direction is to the east. Any deviation from that direction results in a restoring effect that brings the orientation field back into alignment with the electric field. (It's this control of the molecules' orientation that can be used to selectively create patterns on an LCD screen.) If we rotate the vial, the restoring effect will pull the molecules' orientation back toward eastward (bottom right of Fig. 50), and a standing wave, in which *all* the molecules in the material vibrate together around the equilibrium orientation, may ensue. This standing wave of the orientation field extends through the entire medium, in all three dimensions, much as the standing wave at the right of Fig. 48 extends along the full length of the string.[9]

Importantly, the electric field stiffens the orientation field but does not stiffen the medium itself. The medium remains a liquid; the molecules are still free to change their positions. Meanwhile, some of the medium's other fields, such as its flow field and pressure field, remain floppy. As a stiffening agent, the electric field is selective; one field interacts with it and is stiffened, while others are left unaffected. The Higgs field, too, interacts with some fields and stiffens them, but not with others, which it leaves unchanged. (What impact it might have on the everywhere-media that underlie those fields, assuming those media even exist, is something current experiments can't hope to address.)[10]

This last example brings us perhaps as close as we can get to the Higgs field without additional abstractions and a much longer discussion. There are still noteworthy differences. First, unlike the electric field, the Higgs field does not point. This is a matter of detail, albeit an important one for preserving the relativity principle and the coasting law, as I emphasized in Chapter 19. Second, this example involves an ordinary field of an ordinary medium. The lessons learned here may not apply wholesale in more exotic contexts involving amotional media or fermionic elementary fields. Finally, these examples using (relatively) familiar objects provide insight by giving us a way to visualize them. We have no corresponding visual image or conceptual

framework for the Higgs field's influence on other fields; our current understanding of the elementary fields is far too meager. Although these examples
share many surface features as well as math with the case of the Higgs field,
there's no reason to expect them to give insight into the fundamental processes by which the Higgs field does its stiffening.[11]

From these imperfect analogies, we can't draw concrete conclusions, but
we can learn two abstract lessons crucial for the Higgs field. First, a stiff
field, unlike a floppy field, can have broad standing waves—waves that can
extend for great distances and yet vibrate with high frequencies. This explains how a stiff electron field can have standing wavicles. Second, a floppy
field can be made stiff by another field acting as a stiffening agent. This explains how the Higgs field can give the electron field its stiffness. Together,
these factors guarantee the existence of electrons with rest mass.

These basic lessons, when combined with a variety of subtle technical
insights, were enough for physicists of the latter half of the twentieth century to guess what was going on. They managed to develop formulas that
describe the stiffening caused by the Higgs field and predict its implications.
Their formulas taught them how they could search for the Higgs boson and
how to design the LHC so that it would give them the best possible chance
of discovering it. Even today, those formulas are still in use, as experimental
particle physicists attempt to comprehend the Higgs boson and its field in
ever greater detail.

20.3 Farewell to a Phib

This explanation of the Higgs field is in many senses incomplete,[12] and you
may or may not be satisfied with it. But you must admit that it's far better
than the Higgs phib, whose deficiencies are now even more evident.

For one thing, we now see that *the Higgs field has nothing whatsoever to
do with motion*! It stiffens fields; that's all. Wavicles and rest mass did not
even appear in my examples. The Higgs field affects resonance frequencies,
but it's only through Chapter 17's relation between frequency and rest mass,

obtained by combining quantum physics and Einstein's relativity, that the Higgs field's activities change the rest mass of the electron or of anything else.

Along similar lines, it's misleading to say (as I did quite often earlier in this book) that the Higgs field "gives elementary particles their masses." The word *gives* implies that the Higgs field is helping elementary wavicles out, whereas in fact, it's making their lives tougher. In particular, *the Higgs field does not provide wavicles with their energy-of-being*! It's not a source of energy. It's only a source of stiffness, a creator of restoring effects. By making fields stiffer, it increases the energy-of-being required for their wavicles' existence, so it is more a hindrance than a help.

Once sufficient energy to produce a wavicle is obtained, though, that wavicle's larger energy-of-being means it will have more rest mass—more intransigence to serve as armor against the blows that it will face as it travels the cosmos. In this sense, the Higgs field is a double-edged sword. If you'd like to create and collect top quarks, the Higgs field is an obstruction, dramatically increasing the cost and forcing you to expend prodigious amounts of energy. But if your goal is to assemble a large stash of atoms from which to build planets, water, and humans, then the Higgs field is your best friend; without its restoring effects, electrons would have been as effervescent as photons and no atoms would ever have formed.

A final remark on the Higgs phib. We've seen no sign of a universe-filling soup, snow, or molasses, nor should we have. Fields are properties, not substances, and a field's average value becoming nonzero is not (in general) like a substance filling the cosmos. Ordinary fields and their media already teach us this lesson. An iron block is full of iron atoms whether its magnetization field's average value is zero or nonzero. A shift in the magnetization field represents a change in the alignment of the atoms, not a change in the number of atoms in the block. Similarly, empty space, obviously empty when the gravitational field is zero, remains empty (and merely warped) when the gravitational field is nonzero.

As for the Higgs field, maybe it's a property of an amotional Higgsiferous aether, or maybe it's a Cheshire Cat field with no medium, but either way, it's just *in* empty space, integrated into it somehow, like other cosmic fields. Observation confirms that the nonzero average value of the Higgs field has left empty space just as empty, transparent, permeable, amotional, and nightmarish as it would have been if the Higgs field's value had been zero. The only hint of the Higgs field, prior to the discovery of the Higgs boson, was the stiffness of the other fields and the rest masses of their wavicles. We humans are slow, but we finally took the hint.

$$\equiv\; 21 \;\equiv$$

Basic Unanswered Questions

You might have thought that finding the Higgs boson and verifying the reality of the Higgs field would be the beginning of the end of the story. But despite real progress in understanding how the elementary wavicles of nature get their rest masses, significant questions of all sorts still linger. There are basic factual issues that remain unsettled, conceptual puzzles for which we need experimental guidance, and fundamental conundrums that seem difficult to approach.

Before we start to understand these questions, we need to set a proper context. First, we mustn't assume that the Higgs field is responsible for the stiffness of all stiff fields. In Chapter 20.1, we saw that a ball's position can be intrinsically stiff without a stiffening agent, as in the case of the springs in Fig. 46 (p. 267). Analogously, there may exist elementary fields that are intrinsically stiff and require no stiffening agent. It's also possible that we will someday discover fields that get their stiffness from a separate stiffening agent, as yet undiscovered. These are more than idle considerations, since we already know an elementary field that gets its stiffness in a complicated way: the Higgs field itself!

Aside from the Higgs field, though, experiments support a simple story: all the other known elementary fields with stiffness require a stiffening agent of some sort. However, their frequencies, as reflected in their wavicles' rest masses, vary tremendously, as shown in Table 6 (p. 253). One might well wonder whether such disparity is consistent with the MSM, in which all of these fields rely on one and only one stiffening agent.

Surprisingly, the answer is yes, it is possible that the recently discovered Higgs field is the unique stiffening agent for all the other fields in Table 6. The broad range of rest masses and frequencies must then originate from wide variation in the way that these fields interact with the Higgs field. So let's turn our attention to those interactions.

When we refer to two people interacting, we imply that the behavior of each will affect the behavior of the other. The same is true when we say that two fields interact. The wind and air pressure fields interact; differences in pressure will cause air to flow, while strong winds can cause pressure buildup around a building.

As in social contexts, some interactions are stronger than others. Compared to your average acquaintances, you and your friends interact more strongly; you influence each other to a much greater degree. Similarly, the stronger the interaction between two fields, the more readily they affect one another. Because the electromagnetic field and the electron field interact with moderate strength, atoms readily absorb and emit light; also, electron beams can easily be manipulated using magnetic fields, as they were in the cathode ray tubes of twentieth-century television sets. The potent interaction between the gluon field and quark fields is what imprisons all their wavicles in protons and neutrons. The gravitational and electromagnetic fields interact so weakly that even though the Sun's gravity can deflect light, the phenomenon can be detected only during a total solar eclipse and with an excellent telescope.

In the MSM, the stiffness of each elementary field (excepting the gravitational and Higgs fields, which are special cases) derives from a combination of two ingredients: the first is the strength of its interaction with the Higgs field, and the second is the Higgs field's average value. Since there is only one Higgs field in the MSM, with a single average value, *a diversity of stiffnesses must arise from the first ingredient: a diversity of interactions.* The fields are certainly heterogeneous—their frequencies span forty octaves—so it follows that certain elementary fields interact far more strongly with the Higgs field than others do. (The simple math behind these statements appears in note 6 of Chapter 20.)

For example, the *W* field, *Z* field, and top quark field have wavicles with large rest masses, so they must have powerful interactions with the Higgs field. In fact, the Higgs field's interaction with the top quark field must be approximately 340,000 times stronger than its corresponding interaction with the electron field in order to explain why the top quark field's resonant frequency is larger than that of the electron field, and the top quark's rest mass is larger than the electron's, by the same ratio.[1]

If a field doesn't directly interact with the Higgs field at all, as is the case for the electromagnetic and gluon fields, then the field remains floppy no matter what the Higgs field's value is.[2] That's why the rest masses of the photon and gluon are still zero in our universe.

The two ingredients for stiffness impact the fields' frequencies differently. An increase in the Higgs field's value would simultaneously increase the frequencies of all stiffened fields, as well as the rest masses of all their wavicles, keeping the ratios between any two rest masses the same. (For instance, if the Higgs field's value increased by ten times, both the top quark's and electron's rest masses would become ten times larger, with the ratio of 340,000 between them unaltered.) This would be analogous to tuning all a guitar's strings uniformly to higher pitches (i.e., transposing the guitar) while maintaining all their harmonies.[3]

If instead the Higgs field's average value remained the same but the electron field's interaction with the Higgs field increased tenfold, then the electron field's frequency and the electron's rest mass would grow ten times larger, while all other fields and wavicles would remain the same. (As a result, the top quark's rest mass would now be only 34,000 times larger than the electron's, a tenfold drop in their ratio.) The analogy here would be to retuning a single string on the guitar, leaving all other strings alone; this would change the guitar's harmonies.

21.1 Is This Really the Higgs Field?

Up to now, to keep the presentation simple, I've mostly implied that particle physicists are pretty sure they know what's going on—that they've found *the* Higgs boson and confirmed the existence of *the* Higgs field, an elementary field that stiffens all the other known stiff fields. That's the scenario of the MSM. However, the MSM is just the simplest hypothesis that is consistent with the data from the LHC and from earlier experiments. Maybe it's a good guess, and maybe not; only experimental data can settle the issue. In the rest of this chapter, we'll question the MSM's assumptions.

When the new particle was discovered in 2012, scientists' initial task was to confirm that it really is a Higgs boson. Notice I wrote "*a* Higgs boson"—a wavicle of a stiffening agent—and not "*the* Higgs boson"—the one and only.

To accomplish this, the scientists relied on a simple observation: *the stronger the interaction between two fields, the stronger the interaction between their wavicles.* For example, because the electromagnetic field interacts moderately with the electron field but not with the neutrino fields, photons interact readily with electrons but not at all with neutrinos. The fields that interact most strongly with the Higgs field end up stiffened the most, and so their wavicles should exhibit two related features: first, they should have large rest masses, and second, their interactions with the Higgs boson should be strong. Conversely, wavicles with small or zero rest mass should interact with the Higgs boson weakly or not at all.

These expectations can be confirmed or refuted using data collected at the LHC. Specifically, for each type of wavicle, physicists can try to measure (1) its rest mass and (2) its interaction with the Higgs boson. These two quantities should be directly proportional—the larger the one, the larger the other—if the Higgs field is really the stiffening agent for that wavicle's field. Checking this prediction constitutes what we might call the *interaction test*. If the MSM is correct, every type of wavicle from the electron to the top quark should pass it.

Right from the new particle's discovery in 2012, its properties already resembled those of a stiffening agent's wavicle. By 2013, it was clear that the particle was passing the interaction tests at a qualitative level: using methods I'll describe in a few pages, physicists quickly found that it interacts strongly with top quarks, *W* bosons, and *Z* bosons. They also learned that it has little or no direct interaction with up quarks, down quarks, electrons, photons, or gluons. This convinced the Nobel Prize committee that a wavicle of a stiffening agent—some sort of Higgs boson—had indeed been discovered. They decided to award the 2013 Nobel Prize in Physics to the surviving authors of the two papers that were first to discuss stiffening agents of this type, back in 1964: Peter Higgs himself and François Englert. (Englert's coauthor, Robert Brout, sadly did not live to see the discovery and could not be given the award posthumously. Overlooked for the prize was a slightly later paper with similar ideas by Gerald Guralnik, Carl Hagen, and Thomas Kibble.)

Today the evidence is much stronger, so physicists almost universally refer to the new particle as *a Higgs boson* and to its field as *a Higgs field*. Nevertheless, it may not be the universe's only nonpointing elementary field or only stiffening agent. We can't even be sure it's the only stiffening agent for the known elementary fields, as it is in the MSM. Our exploration of imaginary universes reveals that many possibilities besides the MSM are consistent with the data we have so far. Only future data from the LHC and elsewhere can help us distinguish among them.

Although the MSM is sometimes described as a simple theory, it's actually rather complicated, as you can see from Table 6. As I said earlier, it merely represents *the simplest guess that is still consistent with all known particle physics data*. Nature would have been much simpler if its fields were intrinsically stiff and no Higgs field were needed, but that scenario is clearly inconsistent with decades-old experiments, so we can forget about it. However, it's true that nature would be more complicated if it had two or more stiffening agents or if the Higgs field were composite. So we should think of the MSM as the simplest option still available to us.

The fact that our universe can make do with just one stiffening agent is curious and shouldn't be taken for granted, since it's easy to imagine universes in which more stiffening agents would be necessary. It depends on the universe's details. For example, in a universe identical to ours except that its photons have nonzero rest mass, a second stiffening agent (and second Higgs boson) would be mandatory. Conversely, many imaginable, livable universes would need no stiffening agent at all. In a universe like ours but with no W and Z fields, and thus no weak nuclear force, all the fields could be intrinsically stiff.

But even though a single Higgs field would suffice for our universe to behave as it does, we can't assume that there's only one. Nature isn't always frugal and efficient; for instance, two quark types would be enough to make protons and neutrons and all the atomic nuclei we need for life, and yet there are six types. Only experiment can tell us whether the universe has additional Higgs fields, with their own wavicles.

It would also be premature to assume that the Higgs field and its wavicle are elementary. A Higgs boson, like a proton, might be made from multiple wavicles, in which case the Higgs field is a composite field, constructed from multiple as-yet-unknown fields acting together as a stiffening agent. However, a composite stiffening agent inevitably would come along with additional composite fields, built from the same ingredients but organized differently. These would likely include additional Higgs-like fields as well as stiffer cousins of the top quark field, W field, and Z field. Each of these cousin fields would have its own wavicle, still awaiting discovery.

This reasoning presents the LHC experimenters with a straightforward-sounding task, ideal for the two large general-purpose experiments (named ATLAS and CMS) where Higgs bosons were first identified. Within the data collected by those experiments, physicists can search for other types of Higgs bosons and for cousins of other known wavicles. Finding them would provide evidence against the MSM and would suggest that the Higgs field is not unique, not elementary, or both. So far, however, all such searches have come up empty.

But even without discovering new wavicles, experimenters can investigate the MSM in another way: they can check whether the interaction tests are satisfied, not merely qualitatively but precisely. Such high-precision measurements require careful study of large numbers of Higgs bosons. That's why scientists will soon upgrade the LHC to increase its collision rate.

As of 2023, ATLAS and CMS have carried out the interaction tests with good precision for the W and Z bosons, with lesser precision for the tau and for the top and bottom quarks, and with poor precision for the muon.[4] Meanwhile, there's no sign that photons and gluons have a direct interaction with the Higgs boson, consistent with the fact that they have zero rest mass. Nor has there been any indication of an unexpectedly large interaction of the Higgs boson with the up and down quarks, with the electron, or with neutrinos. So far, everything checks out; there's no evidence against the MSM. We might indeed have found the wavicle of the one and only Higgs field.

However, this evidence is far from airtight. The best interaction tests to date are precise only at the level of 5 percent. For most wavicles, precision is no better than 15 percent, and for many wavicles the comparison can't even be made yet.

Perhaps, as the years go by, ever-improving measurements will show that each type of wavicle passes the interaction test. But if the Higgs field is not as simple as in the MSM, someday they will fail. Then a new set of questions will be raised, challenging and changing our understanding of the universe. Whether and when that change might come, and whether it would be cosmetic or profound, is something only nature knows.

21.2 The Secrets of Particle Decay

The interaction tests require measuring wavicles' rest masses and interactions. For some of these, no special experiments are needed; photons and electrons play such a central role in ordinary life that we know all about them already. But other interactions among wavicles have to be teased out of particle physics experiments. I'll give you a sense now as to how this is done.

Perhaps you have been wondering about all the wavicles I've mentioned that you didn't learn about in chemistry class and that play no role in ordinary materials. Why are we made only from electrons and a few other elementary wavicles? It's all in the last column of Table 6. Most types of elementary wavicles spontaneously disappear—scientists say that they "decay"—in less than a second, making them useless for building ordinary objects. (It was for this very reason that, in order to discover the Higgs boson, we had to make our own at the LHC rather than just searching for them in a pile of sand.) The wavicles that survive longer than a second include the ones we're made of; also long-lived are photons, neutrinos, and gravitons, which are abundant in the universe but too mercurial to be incorporated into material objects.

When a wavicle decays, it isn't disintegrating like a complex machine breaking up into its component parts or like an exploding device producing shrapnel. Nor can it vanish into nothing; energy conservation would not permit it. Instead, decay is a transformation: one field's wavicle is converted to wavicles of other fields. It's another example of dissipation, of the sort that transmutes a guitar string's vibrations into sound waves: one field's vibration energy is passed on to the vibrations of other fields.

The vibration of a guitar string decays gradually, its energy carried off steadily by retreating sound waves. But a wavicle's decay is sudden, as though a wizard strikes the wavicle with a wand, uttering a curse or blessing, and makes it into something new. The reason for the difference is that a wavicle is a wave of smallest possible amplitude. A visibly vibrating guitar string can slowly lose amplitude, but for a wavicle, that's not allowed! Its amplitude either must remain the same or jump to zero. If it jumps, its energy must be immediately transferred to other wavicles.

For this transfer of energy to occur, the decaying wavicle needs to interact briefly with the wavicles produced in the decay. As wavicles are just ripples in corresponding fields, an interaction among wavicles can occur only if their fields can interact. Analogously, if a guitar string couldn't interact with the air (or, in field-centric language, with the wind field), its vibrations couldn't produce sound waves. Thus, by studying how a wavicle decays, we learn how its field interacts with other fields. For example, from the fact that a Z boson can decay to neutrinos, we learn that the Z field and neutrino fields interact. Furthermore, the rate at which each decay process occurs, or the probability that it occurs, can serve to measure an interaction's strength.

Conveniently, a wavicle's rest mass can also be measured when it decays, as long as one can detect all the longer-lasting wavicles that emerge in that decay. By carefully measuring those wavicles' energies and their directions of travel, physicists can work backward and infer the initial wavicle's energy-of-being. That's how most wavicles' rest masses are actually measured.

Decays, then, can help experimentalists compile a list of wavicles' rest masses and interactions. They're not quite enough, though, because some interactions don't lead to decays. One reason is that decays have to satisfy a *rule of decreasing rest mass*: the wavicles produced in a decay must have less rest mass in total than the original decaying wavicle. More concretely, this means that wavicles in Table 6 can only decay to wavicles that lie lower in the table.[5]

Other rules can prevent decays, too. Electric charge, a measure of how strongly a wavicle interacts with the electric field, is subject to a rule: in any physical process, the total electric charge of all wavicles involved is conserved (i.e., unchanged). Electrons have electric charge, but because the wavicles below them in Table 6—neutrinos, photons, gluons, and gravitons—do not, an electron cannot decay. If it tried, electric charge could not be conserved; none of the wavicles that could appear in the electron's decay could inherit its electric charge. This gives electrons permanent stability and makes them suitable as ingredients for material objects from atoms to planets. Similarly, because (as noted in Chapter 6.4) the number of quarks minus the number of anti-quarks is conserved, up and down quarks are long-lasting, too. None of the wavicles below them in Table 6 are quarks, so they have nothing to decay to.[6]

It's the combination of these conservation laws and the rule of decreasing rest mass that assures that wavicles with small rest mass are long-lived. By contrast, wavicles with larger rest mass can satisfy the rules more easily. That's why all of them, from muons to top quarks, can decay rapidly, as seen in Table 6.

Ordinary objects that last days and years must be built from durable wavicles. But long-lived wavicles must have small rest masses, which requires that they have very weak interactions with the Higgs field. Consequently, ordinary objects hardly affect the Higgs field, and it hardly affects them—except by stiffening their wavicles' fields. This explains why the Higgs force is completely irrelevant to daily life; its effects are beyond tiny.[7]

At the LHC, Higgs bosons' decays give experimenters many opportunities to measure the Higgs field's interactions with other fields. (Just to be absolutely clear, Higgs bosons, the Higgs field's wavicles, can decay at the LHC, but the Higgs field itself does not decay! Its average value remains fixed and does not dissipate away.) For instance, Higgs bosons often decay to a bottom quark and a bottom anti-quark, implying that the Higgs and bottom quark fields interact with moderate strength. This is consistent with the interaction test for the bottom quark, whose rest mass, roughly eight thousand times larger than an electron's, is indeed moderately large.

You might expect even more Higgs bosons to decay to top quarks, which have much larger rest masses and should interact more strongly with the Higgs boson. But such a decay is forbidden by the rule of decreasing rest mass (see Table 6). Nevertheless, physicists have a second way to measure the top quark field's interaction with the Higgs field. About one in four hundred Higgs bosons decays to two photons through a process in which the top quark field plays a role. Even though the Higgs field and the electromagnetic field do not interact directly (that's why the photon's rest mass is zero), they interact *indirectly* because the top quark field interacts with both of them. (Indirect interactions of this type rely on the quantum uncertainty of the top quark field and are possible only in a universe with a cosmic certainty limit.) From the fraction of Higgs bosons that decay this way, scientists can infer the strength of the interaction between the Higgs and top quark fields. The result passes the interaction test to within 15 percent.

The electron's small rest mass implies an exceedingly small interaction of the electron field with the Higgs field. For this reason, decays of Higgs bosons to electrons and positrons are expected to be too rare to be observed, and indeed, none have been detected.

Though decays make wavicles disappear, the same sort of metamorphosis can be run in reverse. In collisions, wavicles can sometimes be created from scratch, providing additional opportunities for measuring fields' interactions. Two wavicles collide head-on, a magic wand touches the collision point, and—ta-da! Higgs boson![8]

This is the secret behind experimental facilities like the LHC. Smashing particles isn't about dissection or destruction; it's about creation. Imagine slamming rocks together in hopes of making a silver watch or a humpback whale. In a sense, that's what particle physicists do, seeking to generate something new from something old. It works because we're studying a quasi-musical instrument with interacting components; traveling waves in one part of the instrument can be combined to create higher-frequency resonant vibrations in another part.

Within the LHC, where protons collide by the millions every second, short-lived wavicles found neither in protons nor traveling the cosmos are regularly appearing and decaying away. These include top, bottom, charm, and strange quarks; W and Z bosons; taus and Higgs bosons. There might be other types of wavicles, too, ones that we haven't yet been clever enough to discover in the LHC's vast datasets. Though no one can photograph or otherwise directly study these evanescent wavicles, their properties can be inferred from the flying debris in their decays. Relying on these basic techniques of particle physics, scientists continue to search for unknown particles while carrying out the interaction tests, as they seek to learn whether the Higgs field is unique, elementary, and as simple as could be.

— 22 —

Deeper Conceptual Questions

Even if the MSM is correct and the recently discovered Higgs field is unique and elementary, the interactions of the Higgs field pose many puzzles. Some of these go back decades, while others are fresh from the LHC.

22.1 A World Out of Tune

I've sought to entertain you with a poetic analogy comparing the universe to a musical instrument. But I haven't mentioned its harmonies.

One day, after I had taught a class about the relation between mass and frequency, two students came up to chat with me as I walked out of the building. One of them asked if there was something, in the context of the universe, analogous to strumming a guitar.

If you've held a guitar, you know how natural it feels to strum the instrument—to run your hand across it, setting all its strings vibrating. It's almost second nature. If the guitar is in tune, its strings' frequencies are related by simple fractions. For reasons as yet unknown, a combination of notes—a *chord*—with simply related frequencies tends to please the human brain. That's why strumming a guitar that's in tune produces a sound that most humans enjoy.[1]

We chatted about the question. First, even if you could make all the universe's elementary fields vibrate simultaneously, you wouldn't hear anything because none of them make sound waves. And second, unlike a guitar's strings, which typically exhibit standing waves with fixed frequencies, the universe's far-reaching fields can easily have traveling waves with almost any frequencies. But if we limit ourselves to the universe's standing waves, and we convert their corresponding frequencies into sound waves (and slow them down, proportionately, to put them into the range of human hearing), then we can ask the question the students really wanted the answer to: *What is the secret chord—the underlying harmony—of the universe?*

I sat down on a low stone wall and opened my computer. "I won't try to cover the full harmony of the fields; it would take too long, and it wouldn't fit inside human audible range anyway.[2] To keep it simple, let's take three related notes. The electron field, muon field, and tau field are close cousins, identical in every respect except for different resonant frequencies. I'll program in those frequencies, slowed down to audible range, and you can hear what they sound like when played all together."[3]

After typing for a couple of minutes, I sat back and looked at them. "Ready?"

I pressed the Enter key, and the computer produced a ghastly chord, suitable for waking the dead. It startled a couple walking by.

Had the masses of the three wavicles been in simple ratios, like the frequencies of the strings of a well-tuned guitar (or like the frequencies of a major chord, which are in ratio 4 to 5 to 6), the result would have pleased the human ear. But the students were clearly not pleased. One looked as though she'd swallowed a frog.

"Yikes!" cried the other, recoiling. "So the universe is totally out of tune?"

"I'm afraid so."[4]

"Well, compared to our composition faculty's concert last week," said the first student, "it's not really worse. But shouldn't the universe be more beautiful than this?"

I laughed. "Well, you're revealing a philosophical bias! You share it with many scientists of past and present, including famous ones like Kepler and Newton and Einstein, who deeply believed that the workings of the universe *ought* to be beautiful and elegant. But the thing is, even if that bias is correct, it's not obvious what to apply it to. Many scientists have made embarrassing errors by trying to coerce something random to fit their idea of beauty.[5] There might always be deeper patterns, as yet unknown, where the real beauty may actually lie.

"On top of that, this bias might just be wrong: beauty and elegance may be human values that the universe doesn't share. Nobody really knows what the elementary laws of nature are like yet; there are still far too many unknowns.

"But I'm afraid you won't find loveliness and charm in the universe's resonant frequencies. It seems the universal instrument isn't designed for making music attractive to the human ear."

"So much for the music of the spheres," lamented the second student, shaking his head.

The cosmos rings, but it's probably a good thing you can't hear it. You might feel that it is a great shame, a tragedy even, that the universe disobeys the basic laws of harmony. I won't dissuade you from such feelings, but neither will I encourage them; I see no reason why the universe should be under obligation to human aesthetics.

It's important, though, that you not blame the Higgs field for it. All the Higgs field did was acquire a nonzero average value, one that ensures that the frequencies of many fields aren't zero. The Higgs field didn't determine what those frequencies—and the resulting harmonies—would actually be. That role is played by the strengths of its interactions, whose ultimate origin remains completely unclear.

22.2 Why Such Diversity?

Every one of the twelve fermionic fields interacts with the Higgs field in its own way. The extreme diversity of these interactions contrasts sharply with what's found for other classes of interactions. The gravitational field's interactions with other fields are completely universal, and even those of the other bosonic fields are partly universal. For instance, the electromagnetic field has no interaction with any of the neutrino fields. It interacts with the down, strange, and bottom quark fields with equal strength, interacts twice as strongly with the up, charm, and top quark fields, and interacts three times as strongly with the electron, muon, and tau fields. The gluon field interacts exactly the same way with all six quark fields and about twice as strongly with itself, while not at all with the other fermionic fields. The W and Z fields' interactions with the fermionic fields are only slightly more complicated.[6]

I have not mentioned the fermionic fields' interactions with each other, and that's because there aren't any. Fermionic fields can interact with each other only indirectly, when a bosonic field acts as an intermediary between them.

The twelve known fermionic fields can be organized into three "generations," each generation containing one neutrino field, one electron-like field, and two quark fields. Were there no Higgs field, these three generations would have been identical to one another, bringing elegance and symmetry to the universe. But it's all ruined by the Higgs field's chaotically diverse interactions.

Why aren't the Higgs field's interactions organized into a simpler pattern? Why is the universe a profoundly dissonant instrument, and cosmic harmony, by human standards, just a dream? These mysteries take us even deeper than the Higgs field itself, into dark realms where physicists have not succeeded in gaining a foothold. This is not for lack of trying. Many physicists, myself included, have attempted to guess how this striking disorder could have come about. But none of our many theoretical ideas seems to have exceptional merit, nor has experiment yet assisted us with promising clues.

The overall pattern of the fermionic fields is mysterious, too. Why are these fields organized into generations? Why not seven generations, or just one?

Data proves that there can't be any more generations of the type we're familiar with. Had there been a fourth generation with its own quarks, the properties of the Higgs boson would have been substantially different from what LHC experimenters observe them to be.[7] Though we may someday discover other elementary fermionic fields, they would have to be organized in some other way; moreover, their stiffness cannot originate from our Higgs field and would have to arise from another source.

The bosonic fields, meanwhile, are each associated with a well-understood elementary force of nature. As I noted, the gravitational, electromagnetic, gluon, and W and Z fields have simple patterns of interactions with other fields. Because they serve, respectively, as the intermediaries for the gravitational force, the electromagnetic force, the strong nuclear force, and the weak nuclear force, these forces are also governed by simple patterns, making them universal or partly universal. There are guesses as to where these patterns might come from—the oldest and most mathematically elegant goes by the name "grand unification." But none of these guesses is truly convincing. That's largely because the Higgs field and its unruly interactions make a mess of the most appealing suggestions; indeed, the Higgs force, described in Chapter 19, shows no universality or simplicity of any kind.

Might there be more than five forces? Certainly. It is an experimental question, as there's no known principle that would limit the number. Additional elementary bosonic fields and their forces may yet be found.

Stepping back and considering the catalog of known elementary fields and wavicles as well as all the interactions among them, we find a long list of questions and very few answers. The Higgs field, in particular, makes the set of questions far longer than it would otherwise be, and it answers almost none of them. This is hardly satisfying. It seems that we are far from making sense of this universe.

— 23 —

The Really Big Questions

Now we come to the most mysterious unresolved issue involving the Higgs field. It's one that generates a lot of heat but very little light.

The rest masses of the W boson, Z boson, Higgs boson, and top quark are each more than 100,000 times larger than that of the electron. But they're still remarkably small, as is the Higgs field's average value, compared to something called the *Planck mass*. This quantity was discovered by Max Planck when he first brought the cosmic certainty limit h into science. Nowadays, we think of the Planck mass as setting the rest mass of the smallest possible black hole in our universe. More or less equivalently, it is also the largest rest mass that an elementary wavicle can have without becoming a black hole.

Like so many things in this book, the Planck mass is both huge and tiny. It's the rest mass of 10^{18} protons or 10^{21} electrons, gigantic numbers that nevertheless don't add up to much by human standards—just the rest mass of a typical grain of sand. Still, that corresponds to the energy of three tons of TNT explosive, which would be a lot if it were trapped in a single elementary wavicle. With a rest mass equal to the Planck mass, such a wavicle would form a black hole with a diameter a hundred billion billion times smaller than that of a proton. We call this minuscule size the *Planck length*.[1]

One way we might think about the Planck mass and Planck length is that together they set a *cosmic mass-density limit*. No object can have more than one Planck mass in any volume of one Planck length cubed; any such object will form a black hole before (and often long before) it reaches this mass density. That gives us a trifecta of modern physics limits: on speed, on certainty, and on mass density.

The previous paragraph should be taken with caution. It is based on the assumption that Einstein's theory of gravity remains unmodified at distances ranging all the way from the diameter of an atom down to the Planck length. There's no experimental evidence that this is true. If gravity at ultramicroscopic distances differs from Einstein's expectations, then the cosmic mass-density limit may be lower than I've suggested (because the Planck mass may be smaller and the Planck length longer). Nevertheless, to keep this chapter at a reasonable length, I'll stick with this assumption. If the assumption is wrong, then some of the problems I'll describe below will be moderately less severe, though still pressing.

I've brought all this up for a reason. We know how bad it would be if the Higgs field's average value were zero—there'd be no atoms. But what would go wrong at the other extreme—if the Higgs field's average value grew so large that the rest masses of nature's other wavicles, including top quarks, W bosons, and even electrons, approached the Planck mass?

In our universe, the strength of gravity is paltry. Oh, sure, it breaks our cups and glasses when we drop them, and we've only ourselves to blame when we're foolish enough to challenge gravity by climbing a tree and unlucky enough to lose our grip. And yet, using electric forces to power our muscles, we easily push ourselves out of bed, rise out of a chair, and climb stairs, implicitly thumbing our noses at our immense planet's gravitational pull.

It needn't have been this way, nor need it always remain so. Were the Higgs field's average value to begin rising, the rest masses of electrons would increase. So would those of quarks, and eventually they would come to dominate the rest masses of protons and neutrons, causing them to increase as well. Atoms' rest masses would then grow, too, as would their gravitational masses, inflating the mass and weight of every ordinary object.

Initially this weight gain might just be inconvenient, but at some point, it would be catastrophic. Earth survives because the impenetrability of atoms resists the power of gravity's inward pull. But if atoms' masses grew too mighty, gravity would finally win, and the Earth would collapse under its own weight.

All stars, planets, and other rocks would face a similar demise. To survive, we'd need to escape into deep space. Yet this wouldn't help for long. As the Higgs field's value grew ever larger, even creatures our own size would collapse under the force of our own gravity, unable to resist the pull of our torsos on our fingers and toes and heads. We ourselves would become black holes, our origins forgotten, our intelligence crushed. Even bacteria, atoms, and protons would eventually succumb.

It's truly a dark fate. Fortunately, our universe seems destined to escape it. There is no sign that the Higgs field has strengthened or weakened in the past thirteen-plus billion years; in fact, observations of distant, ancient galaxies indicate that it has been completely stable and constant during that vast stretch of time.

Admittedly, this stability does not preclude a sudden change. We'll return to that possibility later.

There's a Goldilocks quality to our universe. If the Higgs field's average value were zero, there'd be no atoms: a calamity. If the Higgs field's average value were immense, humans would collapse: a catastrophe. But in fact, the Higgs field's value is neither zero nor colossal; instead, it's tiny. Consequently, the top quark's rest mass is a ridiculously small fraction of the Planck mass, an electron's rest mass is even smaller, and life in our universe is possible.

The immense gap between the Planck mass and the known wavicles' puny rest masses is called the "mass hierarchy." It is a fact of our universe, but what is its origin? The lack of a clear explanation for why the hierarchy is so wide, and for how the Higgs field's two dangerous extremes were avoided, is often called the *hierarchy problem* or, better, the *hierarchy puzzle* (since it's not entirely clear that it is a problem that needs solving). To see why it's so puzzling requires wading deeper into quantum physics.

23.1 Out of Control

If you take dogs for a daily walk, you probably put a leash on each one. But if the dogs are big, that could be risky. After all, a leash can pull in both directions. Instead of you controlling the dogs, they might end up controlling you!

The whole point of the Higgs field, as it was introduced into the core of particle physics by Steven Weinberg in 1967 and Abdus Salam in 1968, was to give elementary particles their rest masses by stiffening their fields. Nobody at the time worried about the stiffness of the Higgs field; that was assumed to take care of itself.

But as was appreciated later, stiffness is a complicated matter for a stiffening agent. Unlike the other known fields, the Higgs field isn't intrinsically floppy. It can just start out intrinsically stiff; that is, it can be stiff with or without a stiffening agent. In fact, it can get stiffness from many different sources.

What's really surprising is that it can get stiffness—lots of it—from an entirely unexpected place. It can get it from the fields it's trying to stiffen.

This is a form of feedback. It potentially undermines the whole idea of a stiffening agent. What was supposed to be the stiffener may end up the stiffenee.

Runaway feedback typically leads to vicious cycles and extreme outcomes. This is seen in many contexts; for instance, feedback drives inflationary and deflationary spirals that can cause major economic damage. It was partly to temper these cycles, and prevent extremes, that governmental agencies such as central banks and the US Federal Reserve were introduced. But when it comes to the Higgs field's feedback, we don't know what controls it, if anything.

The two possible extreme outcomes for the Higgs field are illustrated at either edge of Fig. 51. Both leave the Higgs field so stiff that the Higgs boson's rest mass is up near the Planck mass. At the far right of the figure, the Higgs field's average value is zero, eliminating the rest masses of electrons and many other wavicles. At the left edge, the Higgs field's average value is so large that the known elementary wavicles end up with rest masses at or approaching the Planck mass.

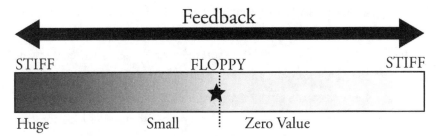

Figure 51: Though feedback from the fields it stiffens pushes the Higgs field toward extremes, it has ended up just left of the center line, as shown by the star, where both the Higgs field's value and stiffness are nonzero but very small. (Not to scale.)

Every field that the Higgs field interacts with produces powerful feedback; the stronger their interaction, the stronger the feedback. This makes the top quark field and the *W* and *Z* boson fields, at the top of Table 6 (p. 253), especially important. Bosonic fields push the Higgs field to the right of Fig. 51, while fermionic fields push it to the left. Adding to the chaos is the Higgs field's own large impact on itself.[2] There could be additional feedback from as-yet-unknown fields if they, too, interact with the Higgs field.

Despite this ferocious shoving match, experiments have shown that the Higgs field has ended up almost exactly in the middle. Its stiffness and average value are nonzero but minuscule. This places the cosmos just barely to the left of center in Fig. 51, as depicted by the star. (Its distance from the center line is greatly exaggerated in the figure.) Somehow, extremes were avoided, disasters circumvented, and here we are. Why?

No one knows. But the mass hierarchy's roots lie here; it is the Higgs field's tiny yet nonzero average value that makes the elementary wavicles' rest masses so minute compared to the Planck mass. As of now, this curious feature of the Higgs field has no simple, experimentally established explanation. Its presence in nature poses what physicists call a *naturalness puzzle*—where the word *natural* has nothing to do with the word *nature*. Instead, it's a synonym for *typical* or *expected*, as in "It is natural for uncontrolled feedback to lead to an extreme outcome, so why didn't that happen here?"

Unfortunately, a full explanation of the feedback's cause is too long to fit into this book. I can only provide you with a brief sketch. The fundamental issue is that fields that are just sitting still in empty space, superficially doing nothing, carry an enormous amount of energy. This is called their *vacuum energy* (because scientists often refer to empty space as *the vacuum*).

The origin of this energy is the cosmic certainty limit—Planck's constant *h*—as applied to quantum fields. In Chapter 17, we touched on how *h* affects "particles." Though we'd expect an ideal particle to have a definite trajectory, with a position and speed at each moment in time, the cosmic certainty limit tells us that quantum particles can't have sharp trajectories (as loosely illustrated in Fig. 40 on p. 238). This can be expressed through Heisenberg's quantum uncertainty principle: it is impossible, at any one moment, to measure or even assign simultaneous meaning to both a wavicle's position and its motion.

When applied to fields, the cosmic certainty limit similarly makes it impossible, at any one moment, to have precise knowledge both of a field's value and of how that value is changing. This has no impact on fields as we encounter them in daily life; for the gravitational field between the Earth and the Moon, the cosmic certainty limit is insignificant, and the same is true for the magnetic fields around ordinary magnets. Nevertheless, this issue has (or seems like it ought to have) dramatic implications for the cosmos.

Way out in the deepest of deep space, far from any stars, stray wavicles, and anything else, we could easily imagine that the elementary fields would be just sitting there, undisturbed, doing nothing. We'd expect, for example, that the electron field would be inert and static; its value would be zero across vast regions of emptiness and would remain so for days, weeks, even years at a time. The amount of energy associated with such inert, static fields would be zero, one would think.

But this can't quite be right. A truly inert and static electron field would violate the cosmic certainty limit. We'd know both the field's value (zero) and how it's changing (not at all).

Instead, the electron field's value has to be somewhat uncertain. This means it can't always be zero; instead, it must always be wavering from moment to moment and from one place to another. (This is related to the zero-point energy mentioned in note 2 of Chapter 16.) All this changing across microscopic distances and times involves energy, and this is what vacuum energy refers to. What's astounding is this: despite the fact that these are microscopic processes, they have very high frequencies, and so the amount of energy associated with them is almost beyond imagination. Even the vacuum energy *density* (the amount of energy in each little piece of space) is huge, potentially reaching the cosmic mass density limit.

This has (or ought to have) a big consequence. Naively, the vacuum energy of just one of the elementary fields ought to be enough to make the universe collapse, or expand, at an eye-watering, mind-boggling rate.

If you look around you, you'll see that this is obviously not happening, so something is off. But please suspend disbelief for a moment and bear with me. The issue that's most important here is that *the amount of a field's vacuum energy depends on its stiffness.* Consequently, any field stiffened by the Higgs field has vacuum energy that depends on the Higgs field's value. This is where we encounter feedback.

Just as the stiffness of a ball's position is a measure of how much effort is required to displace it from equilibrium, the stiffness of the Higgs field is a measure of how hard it is to alter its value. That stiffness is reflected in the Higgs field's resonant frequency and the Higgs boson's rest mass.

In the interest of estimating the Higgs field's stiffness, let's imagine trying to shift its average value. Doing so would change the stiffness of many other fields. That in turn would change their enormous vacuum energies, resulting in a very large adjustment to the universe's total energy. But energy is conserved; this extra energy won't just appear out of the blue. Who is going to supply it? Those of us who want to shift the Higgs field's value will have to pay; we'll have to provide a huge amount of energy to make it happen.

Thus, because the other fields' vacuum energies depend on the Higgs field's value, it takes a tremendous effort to change that value. By definition, this means the Higgs field is inordinately stiff.

This is the feedback I've been referring to. The Higgs field stiffens various other fields, but because those fields' enormous vacuum energy then depends on the Higgs field's value, they in turn stiffen the Higgs field even more than it stiffens them. As a result, the Higgs boson's rest mass—the energy-of-being of a tiny standing wave in which the Higgs field's value vibrates around its average value—would be gigantic by wavicle standards, comparable to the Planck mass. Or so goes the theoretical argument.[3]

The whole problem could be avoided if the other fields didn't interact with the Higgs field in the first place. Then the Higgs field wouldn't affect their vacuum energy and they wouldn't feed back on its stiffness. But that would defeat the purpose; those fields would have remained floppy and their wavicles would still have no rest mass.

Despite what physicists of the 1960s imagined when they invented the idea of the Higgs field, it turns out that the electron field's impact on the Higgs field is potentially much bigger than the other way around. Meanwhile, the electron field's influence is dwarfed by those of the top quark field and the W and Z boson fields. Taken together, the feedback seems destined to push the Higgs field's stiffness to an extreme, all while driving the Higgs field's value either to zero or to an extreme of its own. This would leave the other fields either completely floppy or outrageously stiff.

None of this agrees with experimental data. That's why, when I explain this, I often get skeptical looks from students. "Are you really sure about this feedback?" asked one. "Experiment seems to be suggesting that it's really not there."

"It's a reasonable question," I replied. "Suppose I'm wrong, and the cosmic certainty limit simply doesn't apply to fields. Then there'd be no vacuum energy and no feedback."

Why might one consider that possibility? Well, there's no experimental evidence for a large amount of vacuum energy. The universe's vacuum energy density, which tells us the amount of energy stored in a completely empty box, is a part (and perhaps all) of what is referred to, in shorthand, as "dark energy." (As briefly noted in Chapter 4.1, it also involves a large negative pressure.) Our universe has some dark energy, to be sure. But it's far less than predicted by theorists' formulas, which suggest that the universe ought to have enough energy to destroy itself in an instant. Instead, from the slow pace at which the universe's expansion rate is changing, we learn that theorists are overestimating the energy density of empty space by a trillion trillion trillion trillion trillion trillion trillion trillion trillion trillion, or 10^{120}. (Even if Einstein's view of gravity is completely wrong at ultramicroscopic distances, straightforward quantum field theory would still claim that we're overestimating the vacuum energy density by 10^{40}.) This is known as the *cosmological constant problem*, and it is clearly the biggest mistake ever made in scientific history.

"Gosh, after a failure that bad," remarked another student, "why would you think there's anything *right* about quantum field theory?"

Discarding quantum field theory would seem an easy way out, eliminating both the cosmological constant problem and the hierarchy puzzle.[4] But we would be throwing out the proverbial baby with the bathwater.

"Because quantum field theory's achievements are equally legendary!" I exclaimed. "It has made thousands of successful predictions that rely crucially on applying the cosmic certainty limit to fields. These range from the magnetic properties of the electron to the probability that a Higgs boson will decay into two photons, as tested in delicate experiments that fit on a tabletop and in monster particle accelerators like the LHC. And perhaps the most spectacular success is this: it's the cosmic certainty limit for the gluon field that leads to protons and neutrons! Without it, you and I would not exist."

I'll describe the origin of protons and neutrons in the next chapter. For the moment, my point is this: we cannot escape these puzzles just by blithely abandoning quantum field theory. We'd end up with far more puzzles than we have already.

Not surprisingly, physicists have spent a lot of time thinking about the mass hierarchy. Some have pointed out mechanisms by which the universe could control the feedback. A few have argued that the hierarchy isn't puzzling at all—that the puzzles it poses are philosophical, not scientific. (Personally, I find their arguments lacking in perspective.[5]) One could write a whole book about this subject. Here, I'll just outline the biggest issues.

The simplest possibility is that disaster was avoided *by pure luck*: despite enormous feedback from a wide variety of fields, they all cancel each other out simply by accident. It's like a budget involving multiple unrelated revenues and expenses in the trillions of dollars that miraculously balances to within three pennies. A lucky break of this magnitude is logically possible. I don't view it as particularly plausible, since the cancellation must occur with extreme precision in order to create such an impressive hierarchy. But just because something is implausible doesn't mean it isn't true.

Another obvious possibility is that disaster was avoided *by design*—that our universe was constructed by an external engineer who tinkered with it to assure that the feedback would cancel almost perfectly. This is logically possible, too, though again, is it plausible? One has to wonder why an engineer, in building a universe whose Higgs field has such varied, unruly interactions, would then carefully arrange for the consequent sources of feedback to balance each other to such precision.

Well, who knows? Such speculations may be impossible to check experimentally. Physics is ideally suited for uncovering principles, but both accidents and engineers can be unprincipled. What's a physicist to do?

It may be possible to find some principled evidence in some other way. There's a useful historical analogy. People once thought the number of the Sun's planets and the diameters of their orbits were set by grand principles, and for a long time, they sought to explain them. But no one does so anymore because they're believed to be largely accidental. We know today that many stars have planets, often arranged very differently from our Sun's; moreover, planetary systems can sometimes change, with planets colliding, disintegrating, or being flung out into deep space. There are physics principles that explain how planetary systems form, but no principle requires the existence of our Sun's planets or sets their specific orbits. Maybe there are principles that govern the creation of universes, and once we learn them, they might clarify whether and why the mass hierarchy is unprincipled, perhaps making it less puzzling.

But this could take quite a while. In the meantime, we can try to imagine principled mechanisms that could temper the feedback. We can then try to test each of these ideas experimentally, hoping either to find evidence in its favor or to show that it's wrong or unlikely.

A number of principled mechanisms have been suggested over the years. For example, it could be that the feedback balances automatically: for every fermionic field whose feedback tends to make the Higgs field's average value large, there's a bosonic field with exactly the same strength trying to make it zero, such that the feedback from each pair almost perfectly cancels out. Alternatively, it might be that each field's feedback is far weaker than we might naively think. That would happen if the Higgs field were composite, because the very same forces that could create a composite Higgs field would limit its stiffness. It could also happen if the gravity between objects at extremely short distances, beyond the reach of our best experiments, differs significantly from what Einstein's theory predicts. Yet another possibility is that the history of the universe itself generated a stabilizing effect that controlled the feedback, driving down the Higgs field's stiffness and value and leaving them small and nonzero.[6]

Most of these ideas can be tested experimentally. They predict additional fields that must interact rather strongly with the Higgs field—otherwise they can hardly hope to temper the feedback problem—with wavicles whose rest masses are comparable to or slightly above those of the wavicles we already know. If these additional wavicles exist, they are already being produced at the LHC and may well be discoverable there. On the face of it, it's a golden opportunity: the same particle accelerator suitable for finding and studying the Higgs boson might also be able to give us insight into how the hierarchy problem is resolved.

Unfortunately, there's no guarantee that this will happen.[7] The experimental physicists at the LHC can discover something new and relevant for the hierarchy puzzle only if nature has placed it within reach. This is outside human control; it's entirely dependent on how nature created and sustains the hierarchy.

The LHC's primary purpose was to help physicists find and study Higgs bosons and any of their cousins. In that regard, it has done its job well. Secondarily, scientists hoped it would provide clues to other unresolved questions in particle physics, including the hierarchy puzzle. But as of 2023, the LHC has yet to do so. We have seen nothing unexpected that might clarify the hierarchy's origin or help us address other problems facing particle physicists. Since the LHC was the opportunity of a generation, this has been a significant disappointment.[8]

The story is not over, however. Not only is analysis of existing LHC data (as of 2023) far from complete, but also the LHC, after being upgraded, will collect roughly ten times its current amount of data before physicists shut it down for good.

It's impossible to guess where the saga of the hierarchy puzzle, now in its fourth decade, will take us. It may be four more decades or four millennia before the hierarchy and the Higgs field's interactions are put in proper context, and before those aspects that are based on principles are separated from those that are accidental. In the meantime, I'm sure that theoretical physicists will continue to argue about whether the hierarchy puzzle is really a serious problem, how it might be resolved, and what experiments might be needed to investigate it further. But it's not clear that we'll make any progress until an experiment turns up a clue, next week or next century.

23.2 How Relevant Are These Questions?

In this book, I have focused mainly on twentieth-century discoveries, considering how they influence our understanding of everyday life and our conception of ourselves. Now that you've read about the ongoing mysteries that concern the Higgs field, you might wonder whether these twenty-first-century puzzles offer the potential for equally profound solutions.

Maybe, maybe not. What is so singular about the current moment, when particle physicists have both more answers and more questions than ever, is that anything could happen. The next ten years could bring a revolution, but it's also possible that this book's story won't require substantial revision within our lifetimes. Such a situation hasn't occurred in at least 150 years. Beginning in the 1880s, discoveries in particle physics occurred in every decade, from the first beams of electrons to the interiors of atoms, nuclei, and protons. All along, experiment, theory, and technology were each moving forward, with hints of new insights always lying just around the corner. In the 1930s, the mysteries of the weak nuclear force already pointed toward the collision energy that our current accelerators finally have reached; even back then, one would have correctly expected important discoveries at a machine like the LHC.

But now, in the 2020s, the crisp, clear clues that have pushed particle physics forward for so long have apparently run out. It's simply impossible, at present, to guess what comes next.

This situation is fascinating, albeit frustrating, and it's completely unclear how soon we will emerge from it or how interesting our escape from it will be. Still, there is plenty of potential for substantive new insights. After all, in addition to the long list of questions raised in the last three chapters, earlier confusions still loom in the shadows: the nature of the empty space we travel through, the missing "dark" ingredients of the cosmos, and the origin of the fields out of whose wavicles we're made. Considering the breadth of these mysteries, I doubt that our view of the cosmos has taken final form. It would not surprise me if the future holds another mind-bending revolution of interest to us all.

COSMOS

You might imagine that atomic physicists, nuclear physicists, and particle physicists focus their attention on the microscopic world. But the universe at its smallest influences the universe at its largest, and vice versa. To comprehend stars, galaxies, and the universe's infancy requires knowledge of the cosmic fields and their wavicles. That's why it is typical for a student of particle physics to become well versed in cosmic history, known as *cosmology*, and to learn a certain amount of astronomy.

I've entitled this section "Cosmos," but I could just as well have called it "Quantum II." In these final chapters, I'll explore questions that bring the wider universe into dialogue with quantum physics. How did protons and neutrons come to exist (and why are all protons identical)? How did quantum fields affect the universe's past, and how might they affect its future? And what is the ultimate origin of our bodies' rest mass? After a last look at the relevance of quantum physics to daily life, I'll aim at something that, for any book about science, is truly impossible: a conclusion.

24

Protons and Neutrons

One of the greatest successes of quantum field theory is its explanation of the existence of protons and neutrons. These emerge, in an extraordinarily complex fashion, from the interactions of quark fields and gluon fields. That their properties (and those of their shorter-lived cousins) can be predicted successfully provides unequivocal evidence that fields are subject to the cosmic certainty limit.

Forces between objects typically become weaker as the objects move apart. Naively, we'd expect this behavior, since it would seem odd for two objects separated by half a universe to exert a significant pull on one another. In simple situations, it's true that when a bosonic field acts as an intermediary between elementary wavicles, the resulting force must diminish as the wavicles recede from one another. In fact, its strength will always decrease as fast as the inverse square law seen in both gravity and electromagnetism, or faster, as is the case for the weak nuclear force and the Higgs force.

This would have been true for the strong nuclear force, too, had it not been for quantum physics. The force between two quarks due to the gluon field would itself have satisfied an inverse square law. But quantum physics changes the rules: the gluon field's quantum uncertainty leads to a completely different behavior. David Politzer and his competitors David Gross and Frank Wilczek, who were the first to both calculate and correctly interpret this feature of the strong nuclear force in 1973, were awarded the 2004 Nobel Prize in Physics for this surprising discovery.

The gluon field interacts with itself, and so gluons pull on one another. The strength of the gluon field's self-interaction sets the strengths of all forces between quarks, anti-quarks, and gluons. But this self-interaction creates feedback, far less extreme than the feedback afflicting the Higgs field yet still consequential: it causes the strong nuclear force between wavicles to diminish more slowly with increasing distance than the inverse square law. This trend continues until the wavicles are separated by about a millionth of a billionth of a meter, about the width of a proton or neutron. Beyond that point, the strong nuclear force between wavicles ceases to weaken. It becomes constant and inescapable.[1]

It's no accident that this happens at about the diameter of a proton. This persistent, unrelenting force both makes protons and determines their size, trapping quarks, anti-quarks, and gluons inside them.

All this was learned from theorists' calculations. Admittedly, theorists didn't do so well with the Higgs field's feedback and the cosmological constant. Should we believe them this time?

Yes, because these calculations agree with experiments. They have been checked over and over again, directly and indirectly, through a broad array of measurements and computer simulations. Agreement between theory and data is excellent. There's no getting around it: the predictions of quantum field theory match with nature, confirming that the cosmic certainty limit does indeed apply to fields.[2]

This success provides additional context for the puzzle of the mass hierarchy, clarifying why it is so severe. One can't simply cast scorn on quantum field theory and the feedback it predicts. Instead, one has to explain how theorists could be so right about the feedback that leads to the proton and neutron, along with many other subtle phenomena seen in particle physics experiments, and yet be so wrong about the feedback on the Higgs field.

24.1 The Quantum Cosmic Past

Thanks to all those protons and neutrons, we each carry a powerful nuclear weapon's worth of energy around with us. It's a necessity for the survival of any macroscopic plant or animal. Where did all that energy come from, and how did our parents obtain it?

The energy wasn't poured into us at birth, like gas into a car's tank. We took it in gradually. As a human body grows from a fertilized egg, the number of its atoms continually increases. Since atoms come with rest mass and thus internal energy, a body's energy increases as it grows until a more or less steady state is reached in adulthood. The world has a spectacular amount of energy stored in its atoms, and each of us has borrowed a little, enough to assure that we don't blow away in the wind.

Still, there's a lingering question. The strong nuclear force is responsible for most of the internal energy within a proton, keeping its wavicles inside and assuring that they remain in constant motion. But the strong nuclear force isn't the original source of the energy. It just maintains it, guaranteeing that the proton remains intact and that its rest mass never changes. We haven't yet examined how each proton and neutron obtained its internal energy at the moment of its creation.

The origin story of protons and neutrons begins with the Big Bang phib. *Once upon a time, a singularity, sitting at a point in space, underwent a titanic explosion. A blazing cloud of particles came rushing out into the void, where it cooled, creating the expanding universe we know today.*

Cosmologists—experts on the history of the universe—have already done much to debunk this tale, but nevertheless, it still makes the rounds. I was especially shocked to see it appear in a TV program based on Stephen Hawking's famous book, *A Brief History of Time.*

The truth is more subtle. What we know from observation and theoretical inference is that the universe was once incredibly hot, almost uniformly so. Wavicles swarmed about in a setting more extreme than a proton's belly. You might think that a huge amount of energy such as this would inevitably explode, and that's what the Big Bang refers to. But in fact, nothing exploded during the Big Bang. Imagining it that way is backward.

The Big Bang phib suggests that the universe's birth resembled a bomb detonating in an empty room, creating a fireball that subsequently cooled as it expanded into emptiness. But *the universe had no surrounding emptiness.* Instead, at the earliest times we know anything about, the hot, violent, roiling soup (and in this case, it really was like a soup—an ordinary medium) of elementary wavicles was already everywhere throughout the cosmos. Even possessing the energy of a gazillion bombs, such a thing couldn't explode outward because there was more hot soup blocking the way, also wishing it could explode. It was a universal firestorm, with nowhere to go.

Normally an everywhere-hot soup, unable to cool or spread out, would remain hot and fixed for eternity. So why didn't the universe just stay that way? Because space can stretch. Some unknown event, even earlier in time, gave the universe a kick that caused space itself to expand rapidly.[3] (This kick, or whatever caused it, is what most scientists refer to as "the Big Bang," though not everyone defines the term the same way.) As the space expanded, it gave the wavicles in the hot soup more room to roam, and so the soup expanded, too, allowing it to cool. It's still growing and cooling today, though much more slowly.

Thus, the Big Bang led to an *expansion of space* full of wavicles; it was not an *explosion* that shot wavicles into a preexisting void. This is much more interesting than a bomb. Its detailed origins remain unclear.[4]

Despite this cosmic enlargement, we ourselves are not expanding. Even at the beginning, elementary wavicles didn't grow; only the distances between them did. Protons didn't swell, either; their size is set by the strong nuclear force, and nothing else matters to them. Once elementary wavicles started binding to each other and forming larger units—atoms first, and then galaxies and stars and planets and people—those objects stopped expanding, too. Space continues to stretch, but neither we, the Earth, nor the Milky Way galaxy are inclined to take advantage of there being more room. Not even the cluster of galaxies to which our Milky Way belongs still expands. Today only the distances between clusters of galaxies are still increasing.

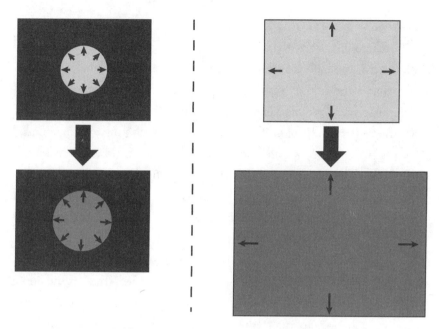

Figure 52: (Left) The Big Bang was not an explosion that blew an expanding hot fireball into a cold void. (Right) Instead, it produced a universal firestorm, hot everywhere; it cooled as space expanded.

Meanwhile, the CMB, the Cosmic Microwave Background that forms the leftover glow from the Big Bang, continues gradually to cool.[5]

But there can't be galaxies or planets without first having protons and neutrons, so let's return to the question of their origin. That brings us to another aspect of the universe's rich history.

We've all boiled water and made ice in the freezer. Freezing and boiling, and the reverse processes of melting and condensation, are all called *phase transitions*, in which an ordinary substance changes its *phase* from liquid to gas or to solid, or vice versa. But there's more in the world than just liquid, solid, and gas. Many materials exhibit less familiar phases. Iron, for instance, has two phases even when solid, one in which it can be magnetized and one in which it can't be.[6]

The universe as a whole can exist in different phases, too. The cooling cosmos has already passed through several, separated by phase transitions. I've already described one of these transitions, in which the Higgs field's average value became nonzero. This occurred a tiny fraction of a second after the Big Bang became hot, when the temperature of the wavicle soup was unimaginable, over a million times hotter than the center of the Sun. At a much lower temperature, closer to that of the Sun's apparent surface, another transition took place. That was when atomic nuclei and electrons, having traveled the universe separately for 400,000 years, began combining to form atoms.

But between these two phase transitions occurred the one of most interest here, after which quarks and gluons found themselves trapped forever. Before this transition, protons and neutrons did not yet exist. Quarks and gluons and anti-quarks roamed freely, forming a cosmic liquid. They dashed around at speeds at or near c, colliding again and again.

Had the strong nuclear force satisfied the inverse square law, this chaos would have gradually calmed as the universe expanded. The cosmic liquid would have rarefied and faded away into a thin atmosphere, its quarks, anti-quarks, and gluons moving farther and farther apart. But instead, once the typical distance between the wavicles reached the size of today's protons, the strong nuclear forces between them stopped weakening and refused to relax their grip.

With the quarks, anti-quarks, and gluons unable to spread any further despite the ever-growing universe, a phase transition occurred. The cosmic liquid began congealing into droplets, each one trapping clusters of quarks, gluons, and anti-quarks inside, along with all of their motion energy.

Most droplets were soon lost. Bigger droplets broke into smaller ones. Collisions among droplets rearranged their quarks and anti-quarks. Droplets lacking extra quarks or anti-quarks quickly decayed to photons, electrons, and other wavicles that aren't susceptible to the strong nuclear force. Yet because of a small surfeit of quarks over anti-quarks (the universe's puzzling matter/antimatter asymmetry, mentioned in Chapter 6.4), a tiny fraction of the droplets with three extra quarks survived. These dissipated their energy as far as they could. In the end, only two types of droplets remained: protons and neutrons.[7]

Now trapped forever inside these droplets, quarks, gluons, and anti-quarks still dash around at speeds at or near c, colliding again and again.

The bedlam of the Big Bang is caught within, never to escape or fade away. The energy in our bodies, and in all ordinary things, is a tiny remnant of the past, a memory of the universe's violent birth.

That's the origin story for our rest mass, and that of all ordinary objects around us. But there's a loose end: Why are all the proton droplets exactly alike? I've explained why electrons are identical—it's characteristic of wavicles—but the logic used there doesn't directly apply to protons and neutrons.

24.2 Dissipation and Quantum Identity

Take any combination of a few wavicles held together by a force, and give them a few moments in quiet conditions. You'll find that the composite particle they form is always the same.

This phenomenon, a consequence of quantum physics and dissipation, explains why all protons are identical. It requires that the number of wavicles be not too large, the wavicles be long-lived, the temperature be low, and certain quantities be conserved. Most atoms at room temperature satisfy these requirements; that's why all oxygen atoms formed from eight electrons, eight protons, and eight neutrons are identical. Snowflakes and other macroscopic objects do not satisfy these conditions, and that's why no two are alike.

The basic idea can be grasped through a simple analogy. Suppose two identical twins on two identical swings are set in motion by their parents, but the parents don't push them in precisely the same way. Then suppose both parents are distracted by their cell phones and stop pushing. The children will be left swinging with different amplitudes, and the timing with which they reach their maximum height will be different. And yet, three minutes later, the twins on their swings will have become identical, as in Fig. 53. Both swings will hang straight down, stationary, the children sitting idle and waiting to be pushed again. The lesson: if there is already underlying identity to build upon, dissipation can enhance sameness.

Quantum dissipation would cause something analogous if you were to pair a million electrons with a million protons. Initially each electron would

Figure 53: (Left) Immediately after their parents stop pushing them, two twins may swing differently. (Right) But dissipation soon brings them to a stop, leaving them identical.

orbit its more intransigent proton in a unique way. But because electrons interact with the electromagnetic field, any electron that is neither stationary nor coasting steadily in a straight line will emit photons. (This is why hot objects glow.) Through this spontaneous radiation of photons, each electron-proton pair would lose energy, causing the electron's motion to simplify and its distance from the proton to shrink. Photon emissions would continue until all available energy had been dissipated. This would leave each electron and proton with the minimal amount of energy that such a pairing can ever have.

All this would happen in less than a second (still long enough for an electron to orbit a proton more often than the Moon has ever orbited the Earth!) Every electron-proton duo would end up in this lowest-energy configuration, called the *ground state of hydrogen*, becoming identical to every other. Though each pair would take a different path along the way, all would share the same destination (Fig. 54), in which the electron wavicle takes on the smallest shape allowed by the cosmic certainty limit. There they would remain, as long as they were kept at temperatures below a few thousand degrees. This is why all hydrogen atoms at room temperature are identical; their sameness emerges automatically thanks to dissipation, without compulsion or labor.[8]

Though more complex than atoms, protons follow the same principle. A proton, in this sense, is simply the ground state—the lowest-energy

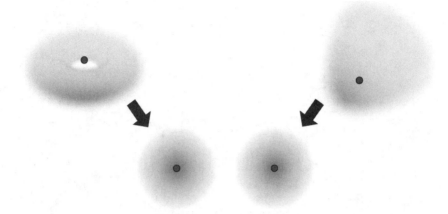

Figure 54: Two electron-proton pairs. (Top) At first they are very different. (Bottom) But after dissipation by emission of photons, they reach their minimal-energy arrangement, the ground state of hydrogen, and become identical.

arrangement—of two up quarks and a down quark. Surprisingly, that lowest-energy arrangement requires a surrounding bath of gluons and quark/anti-quark pairs; even though the energy of that swarm of wavicles is substantial and contributes significantly to the proton's rest mass, removing the bath would cost even more energy. (A quark out of its bath is like a fish out of water: infinitely unhappy.) Similarly, a neutron is the lowest-energy arrangement of one up quark and two down quarks. In far less than a second, a triplet of quarks will end up as either a proton identical to every other proton or a neutron identical to every other neutron.[9]

The larger lesson here is that because wavicles are themselves identical, relatively simple objects built from those wavicles will usually have unique and definite ground states, which they will quickly reach after some dissipation. Any large and complicated system is different. It has so many ways to arrange itself that its lowest-energy configuration cannot be quickly attained, and even if it were to reach that configuration, it would not be able to remain there at room temperature. This is why the objects of the macroscopic world are never exactly alike.

25

The Wizardry of Quantum Fields

A vast, merciless wall sweeps at incredible speed across the heavens. In its path, unsuspecting, lies the universe we know: galaxies, planets, atoms of various elements, stray electrons and photons, and, in a few places, living creatures.

The wall passes in a flash, leaving behind a scene of cosmic ruin. Every star, every planet and asteroid and comet down to the last sand grain—indeed, every individual atom—is instantly obliterated. Death has taken every life form, even the most primitive. Not even subatomic wavicles survive unscathed; many are carried along with the wall, as though it were a giant broom, and most left behind are changed beyond recognition. What was once the dwelling place of the Earth is now a featureless waste, blazing with extreme heat.

The worst is yet to come. As though running the Big Bang in reverse, the fabric we call *space* rapidly contracts. The clouds of wavicles behind the wall are crushed together; the heat and density become exceptional even by cosmic standards. What happens next? We cannot guess. Perhaps space and even time are destroyed, marking the absolute termination of our part of the universe—a true "The End" for everything we know. Or might a new universe rise from the debris of our own? Conceivably, in that newborn universe, life might emerge again, in some distant, unimaginable future....

25.1 The Quantum Cosmic Future

Is this science fiction? We used to think so. We're no longer so sure.

Not that it's a new idea that the universe could suddenly be transformed by a great wall of change hurtling across space, harbinger of a new regime. Scientists have long been familiar with such phase transitions on a much smaller scale.

Take a plastic bottle, fill it with distilled water, and put it in the freezer. After an hour or two, carefully look at the container. You may find that despite having dropped below freezing temperature, the water is still liquid.

Slow and patient cooling of pure water can lead to a surprising outcome. It can fall well below the temperature when ice normally forms and yet remain liquid, as though it had slept right through its freeze appointment. Scientists say that the water is *supercooled*.

But if you want it to stay liquid, be quiet!

Otherwise, tap the bottle or drop a grain of salt into the water. A sufficient disturbance rouses the water from its nap, and then—presto!—somewhere a tiny ice crystal forms, and in the blink of an eye, the crystal grows, its edge a wall of doom sweeping through the water until the entire container is frozen. A mini-universe, swiftly transfigured.

Is it possible that the universe itself is supercooled and potentially subject to a transformative wall of doom? Scientists have long speculated about the possibility. But recent research has confirmed the existence of something capable of causing this type of catastrophe—an apocalypse that no life could survive.

That something is the Higgs field. Ironically, the very field that sustains us today may in the future eradicate our distant descendants. Although the Higgs field's average value is small at present, quantum physics may allow it to jump to a value vastly greater, devastating everything we know.

Just as ice in supercooled water spreads outward from an initial seed, the jump in the Higgs field won't happen everywhere at once. The change will be initiated at a random location, at a random time, through a random and astoundingly rare natural process made possible by quantum uncertainty.[1]

From that location, a wall of doom will rush outward, with the Higgs field's value leaping to extreme levels just behind it. For those living at that time, there will be no escape.

We don't yet know whether this cataclysm, the most violent event since the Big Bang, will ever happen. With towering overconfidence, let's momentarily assume that we humans have already discovered all elementary fields and interactions that involve the Higgs field. If so, then we have almost enough information to infer the fate of the universe. The one thing we still need to know better is the strength of the interaction between the top quark field and the Higgs field. If the interaction is on the stronger side, then the universe is supercooled, but if it's a little weaker, the universe is safe from the wall of doom. Right now, it's a very close call. A difficult precision measurement of the top quark's rest mass will eventually settle the question.

But this is all based on the assumption that the MSM is correct and that there is nothing left to learn about the Higgs field. Such an assumption is dubious at best and difficult to justify, and we will need far more evidence in its favor before we come to trust it. I imagine that decades or centuries will pass before we become confident in our understanding of the universe's future.

Nevertheless, it is thought-provoking. It may well be that the destiny of the cosmos will depend on our stiffening agent, and on the whims and whimsy of quantum physics.

25.2 The Impenetrability of Atoms and the Quantum Touch

After many chapters among atoms and the subatomic world, we've now spent some time in the wider universe. But to close the gaps between the atom, the universe, and the human-scale world of ordinary life, we have to resolve a final puzzle, one that you've been waiting patiently for me to address.

When I described atoms as almost entirely empty, I dodged numerous questions regarding why we don't fall through our chairs, why we don't sink into the Earth, why water and air can be contained within our stomachs and lungs. If the material of our bodies and our surroundings is so intangible, how can we sense the presence of a floor, a book, the hand of another human being? How can anything large and coherent like a rocky planet form and survive? These and a whole host of other questions all boil down to one: "Why are atoms impenetrable to other atoms, even though they are mostly empty space?"

Atoms are not categorically impenetrable. Neutrinos, X-rays, and high-energy protons can (almost always) go right through them. But impenetrability isn't an absolute concept. A chain-link fence allows light through yet keeps humans and other large animals out. Analogously, atoms can let many things go by, but not other atoms.

The fundamental obstruction is that two atoms cannot occupy the same space without the addition of a lot of energy. The reason is again the Pauli Exclusion Principle, discussed briefly in Chapter 17.2. Not only does it determine the structure and chemistry of complex atoms, but it has an equally large role in shaping our everyday world.

The origin of this principle lies in the nature of fermionic fields. I've already emphasized that bosonic fields can have waves of large amplitude while fermionic fields cannot. The reason essentially boils down to a difference in arithmetic.

All waves that we know from daily experience are those of bosonic fields. They and their wavicles satisfy ordinary math. For instance, take one bosonic wavicle of some particular frequency. Now place on top of it another wavicle of the same field, with exactly the same frequency and motion; in other words, arrange that the two identical wavicles are doing exactly the same thing in the same place and time. What results is a wave that resembles a single wavicle, but with larger amplitude and twice as much energy. Next, to your two wavicles moving in synchrony, add a third. The wave again looks the same except that its amplitude and energy are further enhanced. Keep going, and the number and energy of the wavicles follows ordinary math: $1 + 1 = 2$, $2 + 1 = 3$, $3 + 1 = 4$, etc. Eventually you'll have a wave with so many synchronized wavicles that its amplitude will be easy to observe. Do this with photons and you'll end up with a laser. Do this with gravitons and you'll make a gravitational wave big enough for the LIGO experiment to detect.

But for fermionic fields like the electron field, it's different, because their math is alien. Put one electron in front of you. Now try to put another, with exactly the same frequency,[2] on top of the first one. You can't. They'll move apart, or one will have more energy than the other. There's nothing you can do; they refuse to behave exactly the same way. Synchrony is not allowed, because for identical fermions, one plus one isn't two. It's *zero*.

This fermionic arithmetic underlies the Pauli Exclusion Principle. If you try to make two electrons act in concert, the electron field's fermionic math will stop you cold. It's not just that it's hard to make two identical fermionic wavicles do exactly the same thing; it's impossible, because the very idea has no meaning.

This has crucial importance for the existence of life because of its repercussions not only for individual atoms but also for collections of atoms. Without this principle, atoms could interpenetrate, and any objects made from large numbers of atoms would collapse.

That this is true was only fully established in 1975 in the work of Elliott Lieb and Walter Thirring, among others. They significantly improved an initial weak proof by Freeman Dyson and Andrew Lenard from 1967, relying on techniques developed by Fermi and by Llewellyn Thomas in the 1930s. The details are complex; here I will give you the basic idea through a heuristic argument.

As I described in Chapter 17.2, assembling an atom's electrons is roughly like assigning people to chairs in a sloped auditorium, where the rows are filled from lowest to highest in order to minimize the cost in energy. To see the connection with impenetrability, let's take two atoms that are initially far apart; each fills its rows of seats as usual, using the least energy possible. Now, imagine an unrealistic extreme: suppose the two atoms were right on top of each other. Then they'd share a single auditorium; their seats would now be in common. We could take the first atom's electrons and put them in their usual seats. But the second atom's electrons couldn't sit in their usual places because the lowest seats in the auditorium would already be full. Consequently, they would have to take seats in unusually high rows—*and that costs energy*. In short, if distant atoms are to be made coincident, their electrons have to be supplied with extra energy.

More realistically, let's imagine that someone is trying to push the two atoms together; they're not yet on top of each other, but they're getting closer than they usually would. Their auditoriums are starting to overlap, and they're starting to share rows and seats. Well, the argument of the previous paragraph still applies, though only to the electrons on the outer edges. A few of them will have to move to higher seats than usual. The cost in energy is less than if the two atoms were perfectly overlapping, but it's enough that neither you nor I, nor any ordinary process such as dropping a rock from a height or blasting it with dynamite, can squeeze that rock down to an unusually small size.

The Pauli Exclusion Principle isn't the only thing affecting the energy of the two atoms as they approach each other. Within every atom, electrical forces pull electrons toward the nucleus while causing the electrons to repel each other; meanwhile, the cosmic certainty limit keeps the individual electrons spread out. Fermionic math then keeps them even farther apart. As two atoms are pushed together, all of these effects play a role. Which one is really essential for their impenetrability? It is impossible to disentangle them completely. But electrical repulsion is insufficient, as was proven by Dyson in 1967. Bosonic electrons would confront all the same electrical forces and quantum uncertainty as do fermionic electrons. And yet, as Dyson showed, ordinary materials made using bosonic electrons would collapse. It is the math of fermionic quantum fields that keeps atoms impenetrable.

This holds for all atoms. It is impossible to make large numbers of atoms penetrate each other. Even if we obtained enough energy to give it a serious try, our efforts would destroy any object large enough to be a part of daily life.[3]

This is why ordinary material, though full of holes, doesn't feel or act that way.[4] Instead, when you bring your hand near a table, your body discovers that it lacks the energy required to push your hand's atoms through those of the table's surface. Without the needed energy, your hand is unable to move any farther, and this obstacle compresses your skin and causes your nerves to fire. You experience this, colloquially, as a force—created by fermions[5]—that slows your hand to a stop. That, fundamentally, is what it is "to feel the table," or indeed to feel the touch of anything at all.

So when you blink, swallow, breathe, play a guitar, sit back in a chair, attach a leash to your dog, kiss a loved one, or sympathetically put your hand on a friend's arm, think of this: the quantum arithmetic of identical electrons is at work. Without quantum field theory—were electrons not exactly identical, and were they not subject to fermionic math—every object around us would collapse, dissolve, meld with others. There'd be no planet to stand on, and no bodies to stand on it.

=== 26 ===

Coda

The Extraordinary in the Ordinary

No one could deny the importance, to each and every one of us, of Galileo's principle of relativity. While it hides defining features of human existence—Earth's daily spin and its voyage through the void—this principle helps keep our world stable despite all its motion. Without it, the Earth might slow in its orbit, the universe might erode the Earth's atmosphere, and the intrinsic properties of objects might vary dangerously as the Earth rotates daily. The relativity principle helps keep the cosmos a dependable place, one in which a planet can survive and stay warm for billions of years, long enough to host the evolution of self-aware life.

The mind-bending principles of twentieth-century physics might seem far more remote and far less relevant. Einstein's relativity concerns the mysterious, exotic world of superfast things, while quantum physics dominates the mysterious, exotic world of supersmall things. These subjects, so counter to our intuitions, fascinate us. But as slow, lumbering giants, we might seem safely insulated from their bewildering behavior.

Yet this is not so. As we have seen in this book, every aspect of our existence requires them.

That quantum physics matters in daily life is perhaps not so surprising, since large entities are naturally made from small ones. The surprise is more in the breadth of its role. Wavicles, from electrons to photons, are quanta of elementary fields. Protons and neutrons exist because quark and gluon wavicles are imprisoned, trapped by an inescapable force that stems from the gluon field's quantum uncertainty. From these objects, atoms form, protected from collapse by the cosmic certainty limit. The richness of atomic chemistry, on which all biology relies, requires not only the existence of wavicles but also the exotic math of identical fermions: *one plus one is zero*. This unusual math of fermionic quantum fields further assures the stability of macroscopic objects, including ones that act as solids or impermeable membranes, suitable for structures ranging from planets and chairs to skin and the walls of blood vessels. Even in our outsized world, surrounded by objects made from inconceivable numbers of wavicles, our daily activities—walking about, driving a car, reading a book, eating, talking, sleeping—never for a moment escape quantum physics.

The importance of Einstein's relativity to our lives is less obvious, but as we have seen, slow objects with considerable rest mass can be made from fast components whose rest masses are far smaller. Though our protons and neutrons are sluggish, the elementary wavicles inside them are not. Because they travel at or near c, their motion energy far exceeds their internal energy, as does the energy of their imprisonment. Had Newton been correct that an object's mass is its quantity of matter, rather than its quantity of energy, this would have been irrelevant, and we'd be like our Styrenian cousins. But thanks to the relativity formula, protons and neutrons have far more rest mass than do the wavicles that they contain.

Einstein's relativity plays another important role in combination with quantum physics. Together, they create a relation between frequency and rest mass. This relation allows elementary wavicles with rest mass to arise from stiff fields. This in turn transforms the Higgs field, a mere stiffening agent, into an essential ingredient for structure and for life.

Even the Big Bang is far more than ancient history or a spectacular creation myth. It lives on within us. The vast majority of our protons and neutrons were born nearly fourteen billion years ago, formed through the interplay of the expansion of empty space and the quantum uncertainty of the gluon field. These droplets of modified Big Bang fluid got their energy-of-being directly from the ultimate source—and so, therefore, do we.

Of these links between the human world and the universe, what was known when Einstein was born in 1879? Essentially nothing. The empire of Newtonian physics was eroding, but it had not been toppled. The concept of a field had been developed, but subatomic particles hadn't been discovered, much less any relation between particles and fields. Many scientists believed in the existence of atoms, but there was as yet no direct evidence. Unsuspected were cosmic limits on speed, on certainty, on mass density. No one imagined quanta, much less fermionic arithmetic, or a relation between gravity and curved space, or that the universe might be like a musical instrument. The basic underpinnings of ordinary matter, and of the human body in particular, were completely mysterious—and yet no one guessed how truly mysterious they would turn out to be.

The revolutions that overthrew the Newtonian worldview and brought us the one I've tried to convey in this book remain breathtaking even today. Certainly they've transformed science, technology, and society. However, as became clear in the decades after Einstein's death in 1955, their significance reaches even deeper, down to the roots.

It's only in recent years that we've fully appreciated the most important lesson of modern physics: *there's absolutely nothing mundane about ordinary life.* The cosmos, stunningly strange and unrelentingly contrary to common sense, infiltrates our every moment. We ourselves, and everything we experience from birth till death, are vibrating manifestations of a nightmarish, uncertain, amotional universe.

One final fable. Heaven knows I'm no script writer, but let's imagine it as a film.

Ten thousand years from now, after the Earth and humans have been through immense upheaval, warfare, and economic collapse, human societies have resurrected themselves and are working toward the rediscovery of science and technology. They are aided by the recent unearthing of a partially preserved ancient library whose texts they are endeavoring to reconstruct and translate.

The reigning culture differs from ours; its highest scientific priority is the study of Earth's life. While researching marine mammals, whose population is far greater than it is today, scientists notice that whales' songs change substantially when they swim above the Mariana Trench, the oceans' deepest canyon. Specialized submarines are sent to the area to investigate. Several disappear without a trace, but just before they vanish, strange sounds, almost musical, are heard over their communication lines. Finally, one badly damaged sub is recovered. The photographs from its camera reveal no sign of any calamity or of any animal attacking the craft; there's just blurring of the water.

The scientists are baffled. Then one of them, investigating those strange noises, has an epiphany. There are indeed creatures living within the Mariana Trench—the whales are communicating with them—but they're invisible. Rather than flesh and blood, they're built solely *from sound waves in the water*. The submarines weren't recording the sounds the creatures make; those were the sounds the creatures *are*.

They give these wave-beings a name: Ondines. The word (pronounced "on-DEENS" and often spelled "Undines" in English-language texts) comes from the French word *onde*, which means "wave" and is related to *undulate* in English. It refers to sea spirits that have been a part of European mythology since the Renaissance. Though the Ondines of ancient myths merely swim beneath the ocean's waves, the name seems even better suited to these newly discovered creatures, literally made of waves in the water.

It's not long before others wonder whether there might be similar creatures in the atmosphere. These "Ondines of the Air" are soon discovered; made of shrieking ultrasound inaudible to humans, they live in the stratosphere, descending into the clouds at night to feed on thunder. "Ondines of the Earth" are identified shortly thereafter, the source of faint tremors that have long puzzled geologists. Made of low-frequency seismic waves at the base of the Earth's continental crust, they live on the shores of our planet's molten mantle, surviving on volcanism.

The director of the government scientific agency, quoting ancient philosophers (*Earth, Air, Water, Fire*), assembles a task force to search for creatures living within the flames of forest fires. But two young rebels, the stars of the film, are skeptical. Burying themselves in works from the ancient library, they learn from one of the texts that light is itself a wave. They secretly organize a team to search for life made from the waves of daylight. To their great disappointment, they find none. Worse, when word of their failure leaks to the press, they are subjected to widespread ridicule.

Then a friend suggests to them that perhaps luminiferous Ondines can't survive in Earth's atmosphere, with its water vapor, fog, and clouds. Perhaps they should look beyond it. Perhaps the empty space that engulfs the Earth is teeming with life. Imagine cosmic Ondines, giants of light, striding among the stars.

The idea comes as a lightning bolt. For the first time in this culture of the future, which has so concerned itself with the obvious life around it, scientific attention turns to the night sky. To garner support for this astronomical venture, the stars of the film give public presentations entitled "Ondines of the Cosmos?" to all who will listen. "For all we know, the universe is filled with wave-beings," they say in their speeches. "Just think what a joy it would be to discover that the cosmos itself is alive!"

They are persuasive; donations pour in. Telescopes are built on the ancient models. Armed with half-understood texts from the past and aware of the rainbow's secret, their team constructs radio antennas and launches balloons carrying microwave detectors to the edge of outer space. They collect reams of data on electromagnetic waves at many frequencies. Their assistants peruse the data feverishly, searching for signs of organized patterns that might reveal the living universe.

Decades pass with no success. Gray-haired, exhausted, and nearing retirement, they vacation at a beach house, bringing with them the newest translations obtained from the ancient library. Among them is a fragment of a book by an odd name: *Waves in an Impossible Sea*. All that remains is a portion of the middle: a part of the story of fields, along with the entirety of the quantum chapter.

They read it as best they can. As their culture still lacks the math needed to translate the ancient volumes on theoretical physics, they have limited prior knowledge of atoms and none of quanta. They find the text mysterious and difficult to follow. But they proceed, haltingly. Finally, they muddle their way to Einstein's Haiku.

E equals f h,
And E equals m c squared;
From these seeds, the world.

Its sweeping conclusion puzzles them. Wondering if they've missed the point, they go back and read about wavicles one more time. Photons. Electrons. Quarks. *The universe rings everywhere, in every thing.*

And then it dawns on them. All this time, they've been looking in the wrong direction. They don't need telescopes. They don't need antennas or balloons. What they're seeking isn't "out there," and it's not made of light, either.

Ondines of the Quantum Cosmos—*wavicle-creatures*—they're right here. That's what we are. That's what life on Earth has always been.

Acknowledgments

As the final version of this book took form, four people played an especially large role in helping me find its style, strategy, and structure. My literary agent, Toby Mundy, encouraged me to attempt something I had initially tried and rejected: an outwardly conventional popular science book. My partner, Keelia Purscell, pushed in the same direction with similarly good sense and taste and read several drafts, helping me to find a consistent voice and to achieve greater clarity. Together, they gave me the courage to take on larger topics that I had originally sidestepped. I'm deeply grateful to my official editor at Basic Books, Thomas (T.J.) Kelleher, who took a chance on me. His insistence, in our initial discussion, that I write a short statement of the book's inner message helped me take it to a new level, with clearer and more profound goals. In addition to assisting greatly with the book's opening chapters, T.J. made several wise remarks that stuck with me and drove my approach in deciding what to include and what to drop from earlier versions. Meanwhile, my friend Paula Billups, a visual artist and writer by trade, also has formidable editorial skills. She read all this book's major versions, with attention both to detail and to scope. Her advice and criticism have been valuable beyond measure, and I am forever in her debt.

I am pleased to thank several other friends who read and critiqued drafts of the book, among them Sarah Demb, Mark Foskey, Laura Harding, and Dean Simpson. I am similarly grateful to my father for reading the draft at three quite different stages. The final product is markedly better for their questions, insights, and wisdom.

When I first began thinking about this project a decade ago, I had recently started my blog and website, which served as a testing ground for the pedagogical approaches that are found in this book. I cannot overstate how much I have learned, directly and indirectly, from my readers, many of whom asked sharp and pertinent questions, suggested improvements, and pointed out inconsistencies.

The illustrator for the book, Cari Cesarotti, is best known as a theoretical physicist herself. I was taken aback when she asked to be considered for an artistic role in this book, but examples of her work convinced me she was up to the task. I hope you like the results as much as I do!

Many scientific colleagues have taught me the subject of this book, and I am sorry I can't properly thank them all. I am especially grateful to Michael Peskin, who not only was my PhD adviser but also read a late draft and provided numerous helpful suggestions and corrections. He and his Stanford University colleagues Lenny Susskind and Savas Dimopoulos were my most important teachers and mentors in my graduate student days. Other scientists influential in my career include many faculty at Princeton University, Rutgers University, the Institute for Advanced Study, the University of Washington, and my current home, Harvard University. Indeed, the project would not have succeeded without the long-term support of the Harvard faculty in high-energy theoretical physics: Cora Dvorkin, Daniel Jafferis, Howard Georgi, Lisa Randall, Matthew Reece, Matthew Schwartz, Andrew Strominger, Cumrun Vafa, and Xi Yin. I'd also like to thank Jacob Barandes, another physicist and teacher at Harvard, who read and critiqued a late draft of this book.

Finally, with deep sadness, I remember two great physicists, Joseph Polchinski and Ann Nelson, to whom I owe an incalculable debt. Many of my best particle physics papers were written with Ann, and my best string theory papers were written with Joe. These two extraordinary scientists and generous souls were major contributors to high-energy physics throughout their decades of research, and both passed away recently while still in their prime. Their loss has left gaping holes in the community, and in my own heart. It was an extraordinary privilege to learn from them as a student, co-author, and colleague, and this book is dedicated in their memory.

Glossary

amotional: in this book, something amotional always behaves as though stationary; its motion cannot be measured or even defined

amplitude: how far a vibrating object travels as it vibrates (more precisely, half the distance from back to forth)

average value: for a field, its value averaged over a large region of space and time

boson: a particle associated with a bosonic field

bosonic field: a field whose value can be large, whose average value can be nonzero, whose waves can have large amplitude, and whose waves can act in synchrony

composite field: a field known to be made of other fields

composite particle: a particle known to be made of other particles, with a measurable size

conserved: unchanged over time during a physical process; conserved quantities include total energy and electric charge but not rest mass

cosmic certainty limit: in this book, Planck's constant h, which sets the size of quantum uncertainty throughout quantum physics

cosmic field: a field present everywhere in the universe, even inside objects

cosmic mass-density limit: in this book, the maximum density to which any object may be crushed without forming a black hole

Cosmic Microwave Background (CMB): microwave photons making up the leftover glow from the Big Bang; present throughout deep space

cosmic speed limit: about 186,000 miles per second, written as c; the maximum speed of any object with nonzero rest mass and the fixed speed (in empty space) of any object with zero rest mass

deep space: in this book, a region far from any large objects such as stars, planets, and moons; not quite as empty as empty space

elementary field: a cosmic field not known to be composite; may be a fundamental ingredient in the universe

elementary particle: a particle that might not be made from anything else, with a size too small to measure

empty space: a region from which everything has been removed that can possibly be removed

energy: a measure of ongoing activity (as in a moving vehicle) or the capacity for future activity (as stored in the vehicle's fuel); see Table 1, Chapter 8

equilibrium: a location (for an object) or value (for a field) at which the object or field may remain constant and stable over time

everywhere-medium: a medium present everywhere in the universe, even inside objects

fermionic field: a field whose value is always microscopic, whose average value is always zero, whose waves' amplitudes are always minimal, and whose wavicles cannot act in synchrony

field: as summarized in Chapter 15.1; in ordinary settings, a changeable property of an ordinary medium that can be measured at any place or time within the medium; in the cosmos, something analogous that can be measured everywhere and at all times but whose precise origin may not be known

force: in colloquial English and in Newtonian physics, a push or pull of some kind or a class of such pushes/pulls (as in "gravitational force")

frequency: how often a vibrating object travels back and forth and back again

gravitational mass: the property of an object that creates and responds to gravitational forces; in Newtonian physics, the same as mass; in Einsteinian physics, a relative term and not the same as rest mass

Higgs boson: the particle associated with the Higgs field, discovered in 2012

Higgs field: the cosmic field responsible for most elementary particles' rest masses

Higgsiferous aether: this book's name for the Higgs field's (hypothetical) everywhere-medium

internal energy: the amount of energy stored within an object

intransigence: in this book, the tendency of objects to resist change in their motion; quasi-synonym for mass and inertia

intrinsic: independent of the perspective from which an observation is made; opposite of relative

known universe: in this book, the region of the universe that we can potentially observe; same as "visible universe"

luminiferous aether: the (hypothetical) everywhere-medium for the electromagnetic field and for its waves (including visible light)

mass: in Newtonian physics, the same as intransigence and equal to the quantity of matter in an object; after Einstein, ambiguous

mass density: the amount of mass per volume (i.e., the mass of a chunk of material divided by the volume of that chunk)

Minimal Standard Model (MSM): sometimes just "Standard Model"; the particles and their fields known as of 2023, including a single Higgs field and its boson (but usually excluding the gravitational field and gravitons); see Table 6 following Chapter 18

motion energy: the energy an object carries simply because it is moving (or because of other types of ongoing changes that may not actually involve motion); called "kinetic energy" by physicists

ordinary field: a property of an ordinary medium

ordinary medium: a uniform, widespread ordinary material such as water, air, or rock

outer space: usually, any region beyond Earth's atmosphere; sometimes, a region far from large objects

particle: in particle physics, a microscopic object that is a member of a type; all particles of that type are identical

Pauli Exclusion Principle: forbids two identical fermionic wavicles from doing the same thing at the same time, with major implications for atoms and bulk materials

polymotional: in this book, a polymotional object's motion is defined only relative to other objects; the object can be said to have many speeds at once

quantum: the smallest unit of vibration; for a wave, the ripple of smallest possible amplitude

relative: dependent on the perspective from which an observation is made

relativistic mass: a relative property in Einsteinian physics; the intransigence measured by an observer who sees an object as moving and pushes it along its direction of travel; somewhat ill-defined

resonance: here, the tendency of objects to vibrate at their resonant frequency

resonant frequency: the natural frequency at which a vibrating object will vibrate when disturbed and then left alone

rest mass: in Einsteinian physics, an intrinsic property of an object, namely, its intransigence as measured by an observer who is initially stationary relative to the object; equal to the object's internal energy divided by c^2

restoring effect: a force or other effect that causes an object or property to return toward equilibrium, leading it to vibrate (as in a spring or swing)

simple wave: a series of wave crests and troughs of comparable amplitude; see Fig. 22 (p. 141)

standing wave: a wave whose crests and troughs remain stationary, each crest becoming a trough and then a crest again; see Fig. 25 (p. 145)

stiffening agent: in this book, a field that causes other fields to become stiff, after which they can vibrate with standing waves

stored energy: energy within an object that can potentially be turned into (for example) motion energy.

total energy: all forms of energy associated with an object or system of objects; in this book, usually a single object's internal energy plus its motion energy

traveling wave: a wave whose crests and troughs move in tandem from one location to another; see Fig. 23 (p. 143)

value: the amount or strength of a field at a particular place and time

visible universe: in this book, the region of the universe that we can potentially observe; same as "known universe"

wavicle: a quantum of a wave in a medium or a field; often a synonym for *particle* in particle physics because of its particulate nature, though it also has a frequency and an amplitude

weight: the amount of gravity's pull on an object; depends on the object's location relative to other objects

Notes

Chapter 1: Overture

1. As most recently measured in "The Proper Motion of Sagittarius A*. III. The Case for a Supermassive Black Hole," M. J. Reid and A. Brunthaler, *Astrophysical Journal* 892, no. 1 (2020).
2. Many Einstein quotations are misattributions or in error. This one is a paraphrase, by classical music composer Roger Sessions, of something Einstein wrote that had a different meaning.*

Part I: Motion

1. The shadow that the Earth casts on the Moon during a lunar eclipse is always disk-shaped, no matter the time of day, which can be true only for a spherical planet. Earth's size is revealed by comparing the lengths of shadows of two identical objects, separated by a known north-south distance, measured at noon on the same day.*
2. So large is the galaxy that it takes us about 250 million years to complete one orbit.

Chapter 2: Relativity

1. Exactly what constitutes "isolated" is a complex issue if we look at it closely. But informally, an isolated bubble should shield anyone inside from all information about objects outside; otherwise, those objects might create effects inside the bubble that would obscure the relativity principle.*
2. As pointed out by the nineteenth-century French physicist Léon Foucault, the Earth's rotation, the least steady of our motions, is reflected in the motion of a tall pendulum. Many science museums around the world have such a "Foucault pendulum" on exhibit.*
3. The complete story of what we do and don't feel is a bit more complex, involving an interplay of what I've written here with our experience of Earth's gravity and Earth's shape. But these details, while interesting, are not essential in this chapter.

4. Stephen Leacock, "Gertrude the Governess," *Nonsense Novels* (1911). Perhaps Albert Einstein and Werner Heisenberg read it, too.
5. As it navigates, a spaceship may also adjust its path by taking advantage of the tug of gravity from nearby stars, planets, or moons.
6. Were this false, no rocket to the Moon or the other planets could have completed the journey; fuel consumption would have been prohibitively high.
7. This story is confirmed by telescopic images of newly born stars, which often have disks of dust and baby planets around them, and of old stars, which swell tremendously before they die.
8. Another relative speed of note is the plane's speed relative to the Sun, which determines how quickly the aircraft passes through time zones and how quickly the Sun appears to cross the sky. Flying east, a plane is carried along with the Earth's rotating atmosphere, so it moves rapidly across the sunlit half of the Earth; flying west, counter to the Earth's rotation, can delay sunset for many hours.*

Chapter 3: Coasting

1. Readers familiar with astronomy may wonder about the Cosmic Microwave Background, the diffuse leftover glow from the Big Bang at the universe's birth. One can indeed specify one's motion (though not location) relative to this bath of ancient light. But this bath is no more stationary than anything else, even though it is more widespread. Moreover, it would be absent from an isolated bubble, so its existence leaves Galileo's principle intact. I'll say more about these issues later.*
2. The ground at the equator moves, relative to the Earth's center, at over 1,000 miles per hour; the corresponding speed of cities at midlatitudes, such as New York, is about 700 miles per hour.
3. There is a common misunderstanding that Einstein, in his theory of gravity, claimed that whether the Earth orbits the Sun or the Sun orbits the Earth is just a matter of perspective. This is not so if one speaks carefully about how motion and gravity work.*
4. Strictly speaking, any imperfections in the superconductor would cause a modicum of drag. Also there would be air resistance, so for purer coasting, we might want to take the book and superconductor to the airless Moon. Best of all would be to throw the book into deep space.

Chapter 4: Armor Against the Universe

1. Among possible dark matter particles are axions and dark photons, neither of which would obviously qualify as "matter."*
2. Unfortunately, scientific shorthand seems to contradict me. On many websites, one will find the statement that "the universe is made from matter and energy," and even in scientific contexts, one will read "the universe is 5% ordinary matter, 27% dark matter, and 68% dark energy." But these statements don't mean what they seem to say.*

3. Here's one problematic example. Consider a sealed helium balloon. It contains a fixed quantity of matter: a definite number of helium atoms. Any increase in the Higgs field would increase the mass of each helium atom, too, raising the mass of the balloon without changing its quantity of matter. Since mass can change when the quantity of matter does not, they cannot be the same thing.

4. One may check that paper falls as fast as a book by placing a small slip of paper flat on the book's top; the book blocks the air from creating resistance to the paper's fall, and they will indeed land together.

5. GOES stands for Geostationary Operational Environmental Satellite Network, operated by US government agencies.

6. Confusingly, astronauts orbiting Earth inside nearby space stations appear to float as though weightless. From Newton's perspective, they are not truly weightless; if they were, they'd coast, leaving the Earth's vicinity and moving rapidly into deep space. Instead, they and their spaceship are pulled by gravity into a common orbit around the Earth. Since they travel on the same path as their container and as the camera which films them, they seem and feel weightless. (This subtle issue is turned on its head in Einstein's view of gravity.)*

7. In math: by Newton's second law, the acceleration a caused by a force F pushing on an object with mass m is $a = F/m$. Meanwhile, the gravitational force of the Earth on that object depends on its distance r to the Earth's center as $F = GMm/r^2$, where G is called Newton's constant and M is Earth's mass. Together, these imply that the object's acceleration is $a = GM/r^2$. Because this is *independent of* m *but depends on* r, all objects of any mass accelerate identically at Earth's surface, while a distant object, with larger r, will accelerate more slowly.

8. The balance needn't be perfect. Slightly imperfect balance would make the orbit an ellipse rather than a circle. However, no such balance applies to objects on the Earth's surface; we humans do not orbit the Earth. The Moon, and any artificial satellite, orbits roughly in a circle around the Earth's *center*, while each of us is carried around in a circle around the Earth's *axis*. Unlike the motion of the Moon, our motion is far too slow to balance gravity, and were there no solid ground beneath our feet, we'd fall into the Earth.

9. The simplest trick: if the Moon appears to pass in front of a planet such as Jupiter, this eclipse can be seen from only a part of the Earth's surface. The north-south width of that region is approximately the diameter of the Moon. Combining this information with the Moon's apparent diameter on the sky, one can determine its distance from Earth.*

10. To avoid disaster, the Moon's orbital speed would need to be 40 miles per second, leading it to circle Earth twice a day!*

11. Here, by *effect*, I refer to the acceleration caused by gravity, not the gravitational force itself.

12. To explain why gravity leads to a water bulge on both sides of the Earth is too complex for a footnote, and I'd rather not repeat the most commonly heard explanations, which

are phibs. One can see a hint of the cause as follows: if one drops a water balloon in constant gravity, it will fall as a sphere, whereas if it is pulled more strongly at the bottom than at the top, it will stretch into an oval as it falls.*

13. The notion that the Moon caused the tides was widely suspected, but some dissented (even Galileo, who was misled by the Mediterranean's atypical tides). Because Earth's continents obstruct the oceans' flow, the real-world behavior of tides is intricate; Newton himself did not attempt to get all the details right.

14. Alexander Pope, British writer, upon Newton's death in 1727.

Chapter 5: Enter Einstein

1. Here's the formula: if v is an object's speed, and N is its intransigence measured when one tries to increase v, then the object's rest mass is $N\sqrt{1-(v/c)^2}$.

2. An object's rest mass (as normally defined) is always zero or greater. Objects with rest mass less than zero—with negative intransigence—would make no sense; when pushed to the right, they would move to the left, so friction would accelerate them without limit.

3. The slower speeds result from complex interactions between the swift objects and the materials they're passing through.*

4. Indeed, this effect is called *gravitational lensing*.*

5. A photon's gravitational mass depends on its frequency, which we'll discuss in later chapters. That frequency is perspective-dependent; it will rise if you start moving toward the approaching photon and will fall if you move in the opposite direction.*

6. A lovely example of this approach, penned at the boundary of literature and science, is to be found in astrophysicist Alan Lightman's book *Einstein's Dreams*, which explores fictional worlds with alternate forms of space and time. The technique appears in philosophy, too, as in Søren Kierkegaard's *Fear and Trembling*.

7. The scientific buzzwords that describe this way of thinking are *the renormalization group* and *effective quantum field theory*.

8. Einstein's theory of gravity is amazingly elegant as long as one ignores the puzzle of "dark energy," which would have been easier to do had it been exactly zero, and as long as gravity is a very weak force, as its weakness leads to extremely simple equations. In string theory, Einstein's equations become much more complex, and the elegant simplicity of the math shifts to the level of the strings themselves...perhaps.*

Chapter 6: Worlds Within Worlds

1. Cooking is physics, too, but most physics that occurs in cooking (e.g., boiling) is first taught in chemistry class.

2. Lavoisier, a nobleman, was executed during the French Revolution, his appeal quashed by a judge who stated, "The Republic needs neither scholars nor chemists." Fortunately for France, this point of view did not endure.

3. Actually, there'd be no Earth in the first place, since the chemical elements that make up its rocky interior were forged in even more ancient stellar furnaces.

4. The famous book *Cosmic View: The Universe in 40 Jumps* by Kees Boeke, and the later film *Powers of Ten: A Film Dealing with the Relative Size of Things in the Universe and the Effect of Adding Another Zero*, written and directed by Charles and Ray Eames, consider how the world changes when you magnify it by ten over and over again. The strength of their method is that ten is easy to visualize; the downside is that the number of steps is very large. My strategy here is an attempt to be more efficient.

5. Scientists speak of an *order of magnitude*, meaning a range of about ten. If one says "order-of-magnitude one hundred thousand," one typically means more than 30,000 and less than 300,000, though the edges of the range are fuzzy.

6. That is, $10 \times 10 = 100 = 10^2$, $10 \times 10 \times 10 = 1000 = 10^3$, etc.

7. Visible light's waves have crests and troughs much wider than atoms, so a flash of visible light can no more reflect off a single atom than ocean waves can reflect off a pebble.*

8. Avogadro's number, approximately 6 times 10^{23}, is roughly the number of atoms in a gram of hydrogen; 1,000 grams is a kilogram, a little over 2 pounds.

9. The word *nucleus* really just means "inner part" in Latin, and so the central element of a cell is called the cell's nucleus, just as an atom's core is called the atomic nucleus. In this book, where we will not explore cells in detail, the word *nucleus* is always shorthand for *atomic nucleus*.

10. Later he invented the Geiger counter, used to detect radioactivity.

11. If you already know what anti-quarks are, you may be puzzled or even concerned to find them showing up inside protons. But don't worry, it's all fine; see below.

12. Notice that we do not say that "up quarks are particles and up anti-quarks are antiparticles"; that would be incorrect. They are each other's antiparticles.*

13. The weak nuclear force can interconvert up quarks and down quarks and is responsible for processes that change protons into neutrons (and vice versa), such as occur in stellar furnaces. The strong nuclear force leaves quark type unchanged and thus keeps protons as protons and neutrons as neutrons. Both preserve quark number minus anti-quark number, a conservation law that holds in all experiments so far, though it is expected to fail to a tiny degree.*

14. There's a second reason: it works quite well in predicting certain experimental results, for surprising reasons involving sophisticated math. There may perhaps be a third reason regarding the proton's internal structure.*

15. Similarly, an anti-proton and anti-neutron each consist of three extra anti-quarks within a bath of gluons and quark/anti-quark pairs.

16. It is not at all obvious that a proton could have so much inner activity and yet preserve outward stability. In fact, it is possible only because of quantum physics. Atoms have this property, too; quantum physics allows electrons both to surround a nucleus in a static manner and yet to maintain considerable motion.*

17. The behavior of particles produced in high-energy collisions can reveal evidence of inner structure. The higher the collision energy, the more deeply this method can probe. See Fig. 40 in Chapter 17.

Chapter 7: What Mass Is (and Isn't)

1. In this sense, the relativity formula differs from Newton's most famous formula, $F = ma$. The latter cannot be viewed as a relation between only two quantities because the three quantities F, m, and a (force, mass, and acceleration) can all vary independently.

2. There has been controversy among historians as to the role that Einstein's first wife, Mileva Marić, herself a physics student, might have played in the research that appears in Einstein's 1905 papers. Because our focus here is more on the science than on its human origin, I report here the standard history, with the usual caveat that *all* history is at best an oversimplification of the "truth" and often a significant distortion of it. Those interested in the issue may learn more in *In Albert's Shadow: The Life and Letters of Mileva Marić, Einstein's First Wife*, edited by Milan Popović, or *Einstein's Wife: The Real Story of Mileva Einstein-Marić*, by Allen Esterson and David C. Cassidy.

3. Einstein repeatedly received a grade of 1 in math, the highest possible score, only suddenly to receive a 6 in eighth grade. As historians realized in the 1980s, Einstein's school had flipped its grading system so that 6, previously the lowest possible score, became the top score. He was certainly trained; he had a normal undergraduate physics background, was reading the latest papers on his own and with friends, and was writing his doctoral thesis while working in the patent office, where his strong physics background was an asset. Finally, his research never touched on nuclear physics; although he signed a famous letter to US President Franklin D. Roosevelt warning him of the potential of nuclear weapons, the letter itself was written mainly by a leading nuclear physicist, Leo Szilard. Both Szilard and Einstein knew that Einstein's weighty signature would ensure that the message reached the president's inner circle, though its impact on policy seems to have been limited.

4. I, too, risk contributing to mythmaking here. To keep this book short, I am drastically abridging the complex prehistory of Einstein's ideas, which involved numerous scientists, and the long line of theoretical improvements and experimental tests that followed it. It has even been claimed that Einstein's 1905 argument for his formula is not convincing and that Planck gave the first clear argument in 1907.

5. There's even a popular T-shirt that states "You energy, therefore you matter," or sometimes the other way around.

6. A technical point: even some experts might state that black holes can be constructed "entirely from energy." In my view, this turn of phrase puts math ahead of physics; in any specific context, the energy used to make a black hole must be carried by a physical object or field and cannot exist on its own.

7. This is because the definition of relativistic mass is somewhat inconsistent and ambiguous, a point driven home relentlessly by the physicist Lev Okun.*

Chapter 8: Energy, Mass, and Meaning

1. The physics dialect term for motion energy is *kinetic energy*; it can be generalized to include energy associated with other ongoing changes, such as rapid shifts in a magnetic

or electric field. Stored energy also takes many forms. In some, it is called *potential energy*, an especially problematic false friend: instead of something with the potential to be energy, it refers to energy that, stored for the moment, has the potential to be turned into kinetic energy. This jargon tends to obscure rather than clarify, so I'll avoid it.

2. The meaning of the word *heat* in physics dialect differs somewhat from its meaning in English. This is a minor detail in this book but can be important elsewhere.

3. Not exactly true, as some of the stored energy has become heat.

4. In many modern electric or hybrid cars, fuel consumption is reduced by recapturing some of the car's motion energy during braking, returning it to stored energy within the car's battery.

5. Keeping track of energy and its conservation can become ambiguous in contexts where Einstein's view of gravity is essential; these include black holes, the Big Bang, and the universe as a whole.*

6. It's perhaps surprising that a perspective-dependent form of energy can be conserved; you might expect some observers to see it as conserved and others to see it differently. But nature is clever. Although steadily moving observers, looking at an isolated object or set of objects, will disagree about how much total energy it has, they will all agree that the total energy is constant over time. That this all works out consistently is remarkable.*

7. A proof that the intransigence of a stationary object is proportional to its internal energy requires using Einstein's formulas for relativity, which show that to change an object's speed from zero to v requires adding motion energy that's proportional to the object's internal energy. (More specifically, the required motion energy equals the object's internal energy times a simple function of speed, namely, $1/\sqrt{1 - (v/c)^2} - 1$.)*

8. However, some would complain that this interpretation serves merely to define the concept of relativistic mass, which makes it a tautology rather than a relation with conceptual content.

9. Taking E to be *total* energy and m to be *rest* mass, we view $E = m[c^2]$ as true only for stationary objects. Otherwise $E > m[c^2]$; in words, the total energy of a moving object always exceeds its internal energy (by an amount that can be easily expressed in terms of the quantity called *momentum*).*

Chapter 9: That Most Important of Prisons

1. In a lightly modernized and shortened translation, he writes, "We are led to a more general conclusion: the mass of a body is a measure of its energy content; if the energy changes by L ergs, then the mass changes by $L/9 \times 10^{20}$ grams." This buries the lede. Imagine if he had written, "If the mass changes by 1 gram, the energy changes by 9×10^{20} ergs." That would have driven the point home in a more shocking fashion. (An erg is approximately the motion energy of a mosquito.)

2. On the tour, we saw that there are about 10^{28} atoms in your body, but many of them have ten or more protons and neutrons, which requires us to add another zero to this huge number.

3. Though most such weapons have larger rest mass than a human body, they release only a small fraction of the energy contained within their atoms.

4. Maybe not.

5. They can also be trapped inside other short-lived cousins of protons and neutrons, which are generically called *hadrons* and lend their name to the LHC.*

6. In nuclei, atoms, chemical bonds, and planets around stars, the stored energy that holds the system together is *negative*.*

7. The Higgs field's role in a proton's rest mass, and thus in yours, depends on which question you ask. The standard way to interpret the question, which I've used here, assumes that we leave the strong nuclear force's strength the same while we switch off the Higgs field; then the Higgs field contributes a small fraction of a proton's mass. But in a more subtle interpretation, the Higgs field's role can be substantially larger.*

8. Gravity provides another example of an object with rest mass made from objects that have none. A black hole is formed from objects whose mutual gravitational pull traps them in an embrace too tight for them to escape. In principle, a black hole could be made purely from photons, in which case the black hole's rest mass would stem entirely from the photons' trapped motion energy.*

9. The cause of atoms' growth is something called *the quantum uncertainty principle*, which I'll briefly mention later. It assures that with a smaller rest mass, an electron finds its position in an atom more uncertain, causing it to spread out. This makes the atom larger and the electron easier to dislodge.*

10. The power of the explosion and the temperature reached depend on how much of the electron's lost internal energy is released into the explosion, and that depends in detail on how we turn off the Higgs field. It doesn't matter much; it's deadly no matter what you choose to do.*

11. I'm cheating very slightly here. Depending on exactly how the Higgs field's effect is removed, electrons' rest masses might become zero or might instead merely drop by a factor of several billion. Either way, our atoms would disintegrate even in the void of deep space, ripped apart by the (suddenly ferocious) photons of the Cosmic Microwave Background (CMB).*

Chapter 10: Resonance

1. To be 100 percent clear, a note is what one sings on each of the syllables of the "Happy Birthday" song.

2. Complex objects typically have multiple resonant frequencies. A house, for instance, can vibrate in many ways; the same is true of an atom. We will stick to simple objects.

3. These statements are true as long as the amplitude is small enough. When the amplitude is very large, the independence of frequency and amplitude may fail. For instance, overblowing a flute or a recorder pushes the frequency (and the pitch) up slightly.*

4. These parallels are well-known to physicists, especially those with a musical background. My colleague Stephon Alexander emphasized them in his first book, *The Jazz of Physics*.

5. Less ambitious applications of string theory use it as a tool to gain insight into other important problems without demanding that it serve as the ultimate explanation of the universe.

6. According to string theory, many types of elementary particles may actually be one type of string vibrating in different ways, and the universe's many fields may be different aspects of a single *string field.**

7. This assumes that the guitar is purely acoustic. If the player is using a microphone, amplifier, and speaker, these add energy and enhance the amplitude of the sound waves.

8. In physics, strings and springs often make an appearance, and there's always a risk of saying one when you mean the other. Also, there are swings. . . .

9. In math, the force on a pendulum bob equals mxg/L, where x is the horizontal position of the bob, g is the strength of the gravitational field (expressed as the acceleration it causes), m is the ball's mass, and L is the pendulum's length. The frequency of the pendulum is then $\sqrt{g/L}$. This shows explicitly that gravity, a part of the environment, can change the pendulum's frequency.

10. To be fair, though it is irrelevant to the cosmos and to this book, one *can* effectively shorten the electric field in optical cavities, which are to light what wind instruments and organ pipes are to sound. This is not possible for most of the universe's fields.

11. Some musical instruments have strings that are intended to resonate when others are played, giving the instrument a richer sonority. This involves musical harmonics and is unrelated to what the Higgs field is doing.

Chapter 11: The Waves of Knowing

1. This is often called the string's *fundamental frequency.*

2. With care, you can make this kind of wave in a sink or bathtub; let the water tell you how to do it by watching and feeling it slosh. It's tricky but possible to make more complicated standing waves in water.

Chapter 12: What Ears Can't Hear and Eyes Can't See

1. The rainbow phib is that it contains six colors, or seven colors, or five colors, depending on whom you ask and what country they are from. But in fact, when we divide the rainbow into a finite set of colors, we are using a combination of language and perception to create categories that exist only in our brains. The rainbow itself has an infinite set of frequencies without divisions; in no sense is the set of colors finite.*

2. Any form of light causes heating if absorbed; heat is not uniquely connected with infrared light, despite what some schoolbooks may say. It's true, however, that most warm objects around us, including people, coffee, and engines, are too cool to glow in visible light but glow readily in infrared frequencies.*

3. Actually, there's a fourth, wavelength (the distance between crests), but for both light and sound, wavelength is just speed divided by frequency.

4. Two notes are separated by an octave if one of them has double the frequency of the other. Each additional octave doubles the frequency again. Ten octaves cover a range of 1,024.

5. I don't want the CMB to become a distraction, but I also don't mean to sweep the issue under the rug. I've put a thorough discussion of the CMB and its interplay with relativity, explaining why it doesn't affect this book's lessons, elsewhere.*

6. Communication between astronauts on space walks is accomplished by first converting sound to radio waves, transmitting those waves through empty space, and then converting them back to sound using headphones. The same is true for wireless phones except using microwaves through air.

Chapter 13: Ordinary Fields

1. In some contexts, unmagnetized iron splits the difference. Instead of individual atoms being randomly oriented, the material has many microscopic regions, called *domains*; within each domain, the atoms are aligned, but the domains are randomly oriented. The details are different from what is shown in Fig. 32 (p. 175), but on human scales, the effect is the same.

2. If the pressure field increases, a gas becomes more dense, unlike many liquids or solids. If subjected on one side to localized pressure, a solid may move uniformly, while a gas or liquid will flow in a more complex fashion.

3. You might even discover these obscure fields by accident. For instance, you might observe processes that seemingly cause energy to disappear. That energy must have gone somewhere, and some poking around might reveal that it has been carried off by waves in a previously unknown field. Neutrino fields were discovered in much this way.*

4. The most famous is called the *pion field*, a composite of quark and gluon fields; it plays an important role in the structure of atomic nuclei.

Chapter 14: Elementary Fields

1. This is an oversimplification; in fact, the atoms (and, as we'll see shortly, the magnetization field) precess like spinning tops rather than merely rocking back and forth. However, this detail doesn't affect my main conceptual points, and would also be harder to draw.

2. For lack of space and time, I am greatly oversimplifying both the theory and its history. Most notably, although I state throughout this book that gravity results from the warping of empty space, this is not quite correct; it is really the warping of space and time together, viewed as a single bendable object, that leads to gravity. But such details are very subtle and must be left to another book.

3. Caution: do not confuse gravitational waves, waves of space itself, with *gravity waves*, which are perfectly ordinary waves that are caused by gravity. Ocean waves are an example, as they result from gravity pulling a raised part of the ocean downward.

4. The cause of a gravitational wave can be inferred from careful measurements of how the wave's amplitude and frequency change with time.

5. As noted in Chapter 11, the speed of water waves can vary depending on the water's depth and on the wave's frequency. To avoid complications, we should ensure that the water around the boat has uniform depth and that identical rocks are dropped in an identical manner on all sides.

6. I'm implicitly assuming that the space around you isn't rapidly stretching or contracting while you make the measurement, or at least that you are measuring the speed of waves immediately before or after they pass you. This fine point is crucially important when understanding the expanding universe as a whole, though it plays no role here.*

7. If you maintain a constant acceleration by running a rocket continuously, you may be able to stay just ahead of a light wave, but the moment you take a breather, it will catch you.*

8. That these statements (and others in this chapter) are logically consistent was shown by Einstein, using reasoning supported by math. The math is not so complicated, though too long for an endnote; it's the reasoning that's tricky. In any case, Einstein's claims have all been experimentally verified many times over and are woven into modern technologies.*

9. A subtle but important point: this test of relativity would give the same answer even if the flashlights themselves were moving relative to me inside the bubble. Steady motion of a flashlight relative to the bubble affects the *frequency* of the light waves that it emits, from my perspective, but has no impact on their *speed*. The shift in the light's frequency reveals only my motion relative to the flashlight; it reveals nothing about my motion relative to the aether.*

10. In fact, the CMB is just such a substance.

11. These distortions are often called *time dilation* and *length contraction*.*

12. Said another way, Newton-era physicists assumed that there was a universal clock ticking off time across the universe and that all velocities were observer-dependent. But Einstein guessed that time is observer-dependent and there does exist one universal velocity—the cosmic speed limit.

13. That said, from the perspective of someone who remained at your original location, the distance between you and the light *would* grow faster than *c*. This doesn't contradict Einstein, because the cosmic speed limit applies *to the speed of objects relative to an observer, as measured by that same observer*. From a third person's perspective, the distance between you and the waves you're fleeing is permitted to exceed *c*. In fact, if you simply point two flashlights in opposite directions, their light waves do move apart at twice *c* from your perspective, without complaint from Einstein.

14. Again, the math behind this is not so complex. What is challenging is to understand what the math means and how to use it correctly.*

15. An important subtlety: for traveling waves whose speeds can vary, there is a distinction between the speed of the wave's front edge and the speed at which the wave crests move; they are not the same, and I am always referring to the former here. In jargon, these are called *group velocity* and *phase velocity*.*

16. Lewis Carroll, *Alice's Adventures in Wonderland* (New York and Boston: T. Y. Crowell & Co., 1893).

17. One such book would be my colleague Lisa Randall's *Warped Passages: Unraveling the Mysteries of the Universe's Hidden Dimensions.**

18. This is not, by the way, as crude as simply imagining that our world is someone else's simulation. What quantum physics can do is much more subtle and interesting than that. Sadly, this lies beyond the scope of this book, but the issues are partly covered in my teacher Leonard Susskind's book *The Black Hole War*, in George Musser's *Spooky Action at a Distance*, and in Graham Farmelo's *The Universe Speaks in Numbers.**

19. These appear when fields in the full set of dimensions exhibit standing waves within the microscopic dimensions.*

Chapter 15: Elementary Fields

1. Physicists often refer to the average value of a field as its *vacuum expectation value*.

2. These statements are strictly true only over a few miles and for a short time; since the Earth is spherical and spins, these fields' directions change with time and location.

3. Following Bose's work, Einstein proposed what we now call *Bose-Einstein condensates*, a form of matter whose creation in 1995 was awarded the 2001 Nobel Prize.

4. Fermi, an Italian who emigrated to the United States, made monumental contributions to both experimental and theoretical physics and played a central role in the development of nuclear energy and weapons. The largest particle physics laboratory in the Americas is named Fermilab in his honor.

5. This is not to say that a fermionic field can't have any long-range effects. If it has long-lived traveling waves, those waves can travel great distances and have impacts far away, just as ocean waves can. But as we've seen, bosonic fields can cross gaps between separated objects without using traveling waves to do so.

6. There is a big distinction between a laser (a photon beam) and an electron beam; the former acts like a wave of large amplitude, with all its photons working in perfect synchrony, while the electron beam is a long line of little ripples—individual, independent electrons.

Part V: Quantum

1. Another natural misconception: it might be tempting to imagine that photons are to light waves as water molecules are to ocean waves. As an ocean wave passes, the water and its molecules remain in place, merely rocking in little circles. No individual water molecule travels with the ocean wave. By contrast, a light wave's photons all travel with the wave, moving at the cosmic speed limit. Light's medium is not made from photons; its waves are.

Chapter 16: The Quantum and the Particle

1. This is correct for a vibration but a mild oversimplification in the case of a wave, because the actual minimum amplitude of a wave depends on the wave's shape—for instance, on how many troughs and crests it has. Though I will stick with this manner of speaking for now, we'll turn our focus in the next chapter from a wave's amplitude to a wave's minimum energy, at which point its shape will cease to matter.*

2. Actually, even after the vibration is shut off, there is still some random motion left over; physicists call this *zero-point motion*. It arises from quantum uncertainty, which we'll get to soon.*

3. The term *wavicle* may have originated with Arthur Eddington in his book *The Nature of the Physical World*, itself based on lectures Eddington gave at the University of Edinburgh in 1927. Some writers prefer to say that a photon is *both particle and wave*. This perspective, promoted by Niels Bohr in what is known as the Copenhagen interpretation of quantum physics, is a matter of definition, and I leave it up to you. The important thing is to know what a photon's properties are and what it can and cannot do.

4. If language ever made sense in science, it is this object to which we would have given the name *atom*, after the Greek word for "uncuttable." But atoms were theorized and indirectly discovered before their cuttability was known, and wavicles were discovered still later.

5. The similarity in the fields' names is a historical association of both fields with aspects of electricity, but the two fields are profoundly different in character.*

6. Gravitons, wavicles of the gravitational field, may well exist, and I usually assume that they do. But it may be a long time before this can be confirmed experimentally.*

7. In math, the difference is simple. Some fields are described using complex numbers, and they have two types of wavicles related by complex conjugation. For other fields, real numbers suffice; they have one type of wavicle.*

8. Mistaking wave functions for wavelike particles, and thinking them physical objects, is a common error for students learning atomic physics for the first time. Since a single electron can move around in three-dimensional space, a wave function of a single electron exists in three-dimensional space, too, and so it seems like a real object moving in the space we live in. But a wave function of two electrons already exists in six-dimensional space, because the two electrons have six dimensions' worth of possible positions. For four electrons, the wave function is a wave in twelve-dimensional space. Once we get to quantum field theory, the wave function exists in an infinite dimensional space because a field can take on an infinite variety of wavy shapes.*

Chapter 17: The Mass of a Wavicle

1. Strictly speaking, this is true to within the requirements of the cosmic certainty limit.*

2. More technically, what one looks for is proof that the internals of the proton can themselves start vibrating, like the sloshing of milk in a shaken container; this is called an

excited state of the proton. There are simpler but more subtle methods that can measure the proton's size even more easily.*

3. You may have heard that quantum physics is "spooky," as Einstein put it. That's another aspect of the world in which *h* plays a role. But I'm not going to say anything about it here. There are many good books on the subject, such as Sean Carroll's *Something Deeply Hidden: Quantum Worlds and the Emergence of Spacetime* and George Musser's *Spooky Action at a Distance*.

4. If you send a proton instead, it will interact electrically with the atom's electrons whether they are dots or wavicles. That's why a neutron, unaffected by the electric force, gives a better measure of the atom's emptiness. But defining "emptiness" and "impenetrability" is subtle, as we will see in later chapters.

5. Sadly, a number of scientists exploring X-rays and gamma rays died young because they either did not know or did not appreciate this principle.

6. When a photon is absorbed, the photon strikes the atom and disappears, while its energy is taken up by the atom. Emission of light is the exact reverse; the atom gives up some excess energy and transfers it to a spontaneously created photon.*

7. How broad does this standing wave's crest really need to be to count as a "stationary" electron? For our purposes, a lot wider than an atom but narrower than a human. If it's a millimeter across or more, any associated uncertainty in its motion has slowed below walking speed, which is far less than needed for a precise rest mass measurement.*

Part VI: Higgs

1. Higgs's view has been widely quoted, as in "Prof Peter Higgs: Atheist Scientist Admits He Doesn't Believe in 'God Particle,'" *Telegraph*, April 8, 2013, https://www.telegraph .co.uk/news/science/science-news/9978226/Prof-Peter-Higgs-Atheist-scientist-admits-he -doesnt-believe-in-god-particle.html, where he stated, "I know that name was a kind of joke and not a very good one. I think [Leon Lederman] shouldn't have done that as it's so misleading."

Chapter 19: A Field Like No Other

1. In physics dialect, it's the only elementary field without *spin*, and so the Higgs boson is the only wavicle that doesn't act as though it rotates. The other bosonic fields have spin 1, except the gravitational field, which has spin 2; the elementary fermionic fields have spin ½.*

2. A nonzero electromagnetic field would undermine the principle of relativity in an even more insidious way. As noted in Chapter 14, the magnetic field around a magnet will appear to you, if you rush past it, to be partly electric. The greater your speed relative to the magnet, the more electric field you will detect. Suppose our universe had a constant nonzero magnetic field everywhere, as though we lived within a gigantic magnet. Then by measuring how much of the nonzero electromagnetic field around us seems electric

rather than magnetic, we could gain clues about our motion through the universe. Similar issues would afflict any uniformly nonzero pointing field.

3. In my discussion of pointing and nonpointing fields, I've conflated some conceptual matters and swept a layer of complexity under the rug. This layer, which would have required a further exploration of Einstein's view of relativity, would be needed for a proper comparison of the Higgs field with the electromagnetic field. I've attempted to capture the spirit of the issues in this chapter, even though I can't convey them in their entirety.*

4. This consequence of the Higgs force was studied in my first particle physics paper, written in 1990 with my PhD adviser, Michael Peskin. At the time, I expected the measurement to be carried out by 2010, but it's no longer clear that I will live to see it.

Chapter 20: The Higgs Field in Action

1. As I've told it, this story represents physicists' best guess as to what happened. Though we know a lot about the early universe, the sequence of events involving the Higgs field has not been directly verified via experiment or observation; it is inferred from theorists' formulas.

2. As described in note 9 of Chapter 10, a pendulum's frequency is proportional to $\sqrt{g}$, where g is the strength of the gravitational field. This shows explicitly that gravity acts as a stiffening agent: larger g gives the pendulum a higher frequency, while if g is zero, so is the frequency.

3. Excepting the Higgs field, all known elementary fields are at equilibrium when their value is zero (which is precisely why their average values across the universe are zero).

4. Here, the restoring effect is proportional to kx, where k is the springs' stiffness constant and x is the distance of the ball from its equilibrium point. The stiffness of the ball's position, and its frequency of vibration, is then $\sqrt{k/m}$, where m is the ball's mass. Note that g does not appear; gravity plays no role.

5. For a ball with an intrinsically stiff position, a spring creates a restoring effect kx and a vibrational frequency proportional to $\sqrt{k}$; see note 4 above. Similarly, the stiffness of an intrinsically stiff field Φ results from a restoring effect proportional to $m^2\phi$, where ϕ is the value of the Φ field and m is the rest mass of the Φ field's wavicles. The field's resonant frequency f is then proportional to $\sqrt{m^2} = m$. This is exactly as required by the basic relation between m and f that we learned in Chapter 17.2.

6. As an example of how rest mass arises for a wavicle that gets its mass from a stiffening agent, consider the Z field. Its stiffness is provided by the Higgs field H, which creates a restoring effect proportional to $(y_z v_h)^2 z$, where v_h is the Higgs field's nonzero average value, z is the Z field's value, and y_z is the strength of the interaction between them. The resonant frequency for the Z field and the Z boson's rest mass are then proportional to $\sqrt{(y_z v_h)^2} = y_z v_h$. Similarly, the electron field's frequency and the electron's rest mass are proportional to $y_e v_h$, where y_e is the strength of the interaction between the Higgs and electron fields. The ratio of the Z boson's rest mass to that of the electron is then

simply y_z/y_e. More generally, the rest mass of a wavicle is equal to the product of (1) the strength of the interaction between its field and the Higgs field and (2) the Higgs field's average value.*

7. Intrinsically stiff fields would be symmetric in a mirror. But it turns out that processes involving the weak nuclear force are not symmetric; this was discovered in 1957 by Chien-Shiung Wu, one of the foremost experimenters of her generation, after a proposal by theorists Tsung-Dao Lee and Chen-Ning Yang. (The fact that only the theorists won a Nobel Prize for this achievement is widely considered a historic injustice.) From this it was gradually understood that the known stiff fields need a stiffening agent.*

8. Quantitatively, these waves' frequencies will be somewhat larger than they were without the curtain.*

9. In practice, dissipation in a liquid crystal is so strong that the vibration dies out almost immediately, like the sound of a guitar string in contact with a blanket. But the point here is one of principle.

10. Just as the Higgs field interacts more strongly with some fields than others, the same would be true of the electric field's interaction with two different types of liquid crystals in adjacent vials. Because the molecules of the two liquid crystals have different shapes, the restoring force generated by the electric field on their orientation fields would also differ, and so one's field would be stiffened more than the other's.

11. In string theory, sometimes one can understand a Higgs-like field pictorially. For instance, its average value might represent the physical distance in an extra dimension between two slabs of ordinary space. In other cases, no simple picture is available. I mention this to emphasize that a detailed mechanistic picture of how the Higgs field works might someday be found—or not.

12. Most notably, I have omitted fascinating details about how the Higgs field affects the W and Z bosons and the photon, and thereby the weak nuclear and electromagnetic forces.*

Chapter 21: Basic Unanswered Questions

1. Experimentally, the logic works in exactly the reverse order; we measure the top quark's and electron's rest masses, use our m-to-f formula of Chapter 17.2 to learn what their fields' resonant frequencies are, and deduce what their fields' interactions with the Higgs field must be.

2. The gravitational field, as always, is an exception; though it interacts with the Higgs field, it is not stiffened by it. This reflects the special, universal way in which gravity interacts with everything.*

3. Neutrinos may or may not follow this rule. If the Higgs field's value increases ten times, their rest masses might increase tenfold or perhaps a hundredfold (i.e., by ten squared). Experiments have yet to weigh in on the matter. More generally, experiments are actively trying to settle several important related questions: why neutrinos' rest masses are so small, whether they interact in a unique way with the Higgs field, and whether they

are their own antiparticles. In this book, I have consistently swept neutrinos to the side, not because they are uninteresting but because they are too interesting and would create a distraction.*

4. The data as of 2022 are given in these two papers: "A Detailed Map of Higgs Boson Interactions by the ATLAS Experiment Ten Years after the Discovery," The ATLAS Collaboration, *Nature* 607, nos. 52–59 (2022), and "A Portrait of the Higgs Boson by the CMS Experiment Ten Years after the Discovery," The CMS Collaboration, *Nature* 607, nos. 60–68 (2022).*

5. The origin of this rule is the conservation of energy.*

6. Down quarks and up quarks may decay to one another under particular circumstances—this is why neutrons on their own are unstable—but in many nuclei, such a decay cannot occur.*

7. There's a second reason why it is irrelevant. Because the Higgs field is quite stiff, its force declines very rapidly with distance. The same is true of the W and Z fields; this is what makes the weak nuclear force so weak.

8. The most common process that can create a Higgs boson involves the collision of two gluons, one from each of two colliding protons. The process is generated through an indirect effect in which the top quark field is an intermediary, similar to the one briefly described in the main text that allows Higgs bosons to decay to two photons.*

Chapter 22: Deeper Conceptual Questions

1. The six frequencies on many guitars, corresponding to the notes *E, A, D, G, B, E,* are in proportion 18, 24, 32, 45, 54, 72, or equivalently, relative to the highest-frequency string, $\frac{1}{4}, \frac{1}{3}, \frac{4}{9}, \frac{5}{8}, \frac{3}{4}, 1$.

2. I glossed over other problems. For instance, there's no precise, unambiguous definition of the up, down, and strange quark rest masses. That's because the powerful forces keeping these quarks trapped never allow them to be stationary and isolated.

3. Actually, because the three frequencies are so far apart, I took the liberty of multiplying the electron field's frequency by sixty-four, raising it six octaves. This made the frequencies easier to hear and would have preserved nice harmonies, if there were any.

4. If we put the electron field's frequency on a musical *C*, then the muon field is a very sharp *A♭* seven octaves higher and the tau a sharp *A* eleven octaves higher. We could also include the *W, Z,* and Higgs fields; these would be a sharp *E♭*, a very sharp *F*, and a flat *B*, all seventeen octaves above the electron field. Either triplet of notes, or the six of them together, represents a discordant harmony both within the pure harmonic series and within a piano's even-tempered scale.

5. Among these were Kepler and Newton themselves.*

6. Scientists do understand, from advanced math, why spin 2 and spin 1 pointing fields show universality, while nonpointing fields such as the Higgs field need not do so.*

7. The rate at which the LHC produces Higgs bosons would have far exceeded the rate predicted by the MSM.*

Chapter 23: The Really Big Questions

1. A wavicle with rest mass fifty times the Planck mass would form a black hole whose diameter is fifty times larger than the Planck length, and so on.
2. The Higgs field interacts with itself, and so it, too, feeds back on its stiffness.*
3. This is a significantly oversimplified argument because the energy involved could be both positive and negative, while in my reasoning, I've tacitly assumed it's always positive. But it gives a sense of the scale of the problem.*
4. Actually, the cosmological constant problem is worse than I've made it sound. Even if the fields' vacuum energy were absent, the problem would remain, because effects from cosmic phase transitions (see the next chapter) also contribute to the universe's energy density. But conversely, there is an argument, called the *anthropic principle*, that offers a plausible resolution of the cosmological constant problem. Unfortunately, this argument can't resolve the mass hierarchy puzzle unless one makes powerful and dubious assumptions that create an equally serious challenge, sometimes called the *artificial landscape problem*.*
5. To explain this point of view requires a long, careful discussion of the mass hierarchy and quantum field theory. I have written about it elsewhere.*
6. Buzzwords that go with these classes of suggestions include *supersymmetry, composite Higgs field, large extra dimensions*, and the *relaxion*. There's no experimental evidence for any of them as yet.*
7. Regrettably, some particle physicists stated or implied in public that there was indeed a guarantee. I do not fully understand why they did so.*
8. A further goal of the LHC was to reveal secrets of the strong nuclear force; here, it has been a remarkable success, full of interesting surprises that will someday deserve a book of their own.*

Chapter 24: Protons and Neutrons

1. For the phenomena described in this section, this is essentially true. But if you pull a quark and an anti-quark sufficiently far apart, something more complicated happens. The details would take us too far afield.*
2. A recent high-precision computer simulation was able to calculate the small difference between the proton's rest mass and the neutron's, a shift of just two-tenths of 1 percent. "Ab Initio Calculation of the Neutron-Proton Mass Difference," BMW Collaboration, *Science* 347, no. 1452 (2015).
3. The prevailing best guess is that the kick was created through a phenomenon known as *cosmic inflation*. If this is correct, then the hot soup was created, more or less simultaneously everywhere, when the period of rapid inflation came to an end. This involves even more quantum cosmic physics. See, for example, Alan Guth's book *The Inflationary Universe: The Quest for a New Theory of Cosmic Origins*.*

4. The common statement that "the universe started with a singularity" is speculative. It derives from a theorist's calculation, but the math involved is suspect precisely *because* it shows a math singularity—an infinity. There's no observational evidence for or against there being a real, physical infinity at the universe's beginning.*

5. Now, finally, we can get a full perspective on the CMB. It is an ordinary medium, not an amotional everywhere-medium, as it is absent inside stars, planets, and properly isolated bubbles. We can measure our speed relative to it, and when we do, we learn our speed relative to the original hot soup of wavicles, which was itself an ordinary medium. Do not confuse this with measuring our speed relative to empty space, which is impossible if empty space is amotional.*

6. An ordinary iron magnet melts at 2800 degrees Fahrenheit (1540 degrees Centigrade), but its magnetism will disappear at 1420 degrees Fahrenheit (770 degrees Centigrade).

7. I have not explained why the only acceptable droplets always have three extra quarks rather than some other number. This has to do with details of the strong nuclear force that play no other role in this book.*

8. For all this to work out, it's crucial that hydrogen's ground state is unique and that considerable energy is needed for the atom to jump to any other state. This is a consequence of quantum physics that I do not have time to cover. Had it not been true, identical endpoints for all electron-proton pairs would not have been guaranteed.*

9. How dissipation works with quarks and gluons is more complex than for electrons in atoms, and there are many more interesting details even though the main lesson is the same. For example, other combinations of quark triples (such as three up quarks) have their own ground states, and particle physicists observe and study them, but they exist for less than a second.*

Chapter 25: The Wizardry of Quantum Fields

1. The cause would be a quantum process called *tunneling*, the principle behind a *scanning tunneling microscope*.*

2. Also, to be precise, their spin orientation should be the same. Electrons' spin, though of great importance generally, is merely a distraction here.*

3. Fermionic math has cosmic roles, too, in preventing not only planets but also neutron stars and white dwarf stars from collapsing under their own weight.*

4. It may or may not look that way, depending on how the material interacts with light; for instance, glass looks empty but doesn't feel empty.

5. This puts the lie to the common statement that bosons are "force particles." It's true that forces at a distance are caused by bosonic fields. But most forces that we experience every day are the forces of contact—feet on a floor, keys in a hand, raindrops on an arm—and these involve fermionic math.

Index

Note: Page numbers ending in " f" indicate figures; page numbers ending in "t" indicate tables.

aether, Higgsiferous, 203, 211t, 277
aether, luminiferous, 162–163, 166, 189,
    194–196, 199–202, 204–210, 211t
air pressure, 173, 257–258, 258f, 280
air resistance, 1–3, 24–25, 35–38, 38f, 43–44,
    53, 110–111
airplanes, 16, 24–28, 27f, 28f, 37–39, 38f,
    43–44, 110–111, 153, 195, 198–200, 204
aluminum sheet, 177–178, 178f, 208, 211t,
    240, 249
ambimotional object, 23, 29, 200
amotional medium, 199–203, 206–210, 218,
    219t, 230, 236, 252, 259, 275–278, 329
amplitude, 126–130, 137f, 141–143, 143f, 186,
    223–231, 240–241, 243
antimatter, 92–94, 316
antiparticles, 91–94, 230–231
anti-quarks, 91–95, 92f, 118–119, 230–232,
    253t, 261, 287–289, 312, 316, 319
Aristotle, 36–37
atomic nucleus, 87–95, 88f, 90f, 97–100, 117,
    120–121, 239–240, 284, 316
atoms
    assembling, 124, 244–250, 324–326
    cells and, 83–87
    depictions of, 85, 85f, 87–88, 88f, 90, 90f,
        93, 121f, 319f
    Einstein and, 85, 241
    electrons and, 5, 80–81, 85f, 87–88, 88f, 90f,
        92–95, 97–99, 104–105, 113, 116–121, 121f,
        122–124, 166–167, 215–216, 221–222,
        234–241, 248–250, 277, 287–288,
        295–298, 306, 316–320, 324–329

empty space and, 88–90, 166, 192–194,
    239–240, 323–325
in human body, 86, 241
hydrogen atoms, 49, 84, 318–319
impenetrability of, 89–90, 194, 240, 297,
    322–326
in iron block, 174–176, 175f, 187f, 277
large numbers of, 86, 324–325
matter and, 49–50, 79, 94–98, 113,
    328–329
nucleus of, 87–95, 88f, 90f, 97–100, 117,
    120–121, 239–240, 284, 316
orientation of, 187–188, 187f
oxygen atoms, 248–249, 317
quantum field theory and, 322–326
radioactivity and, 43
size of, 84–86
average value
    of Higgs field, 213–218, 219t, 256–266,
        273–281, 288, 292, 295–305, 299f,
        315–316, 321–322
    nonzero, 213–218, 219t
    zero, 213–218, 219t, 233, 233f, 256–262,
        265–266, 273, 277–278, 288, 292,
        295–302, 299f

Becquerel, Henri, 100
bending field, 174–178, 178f, 183, 208, 210, 211t
Big Bang, 86, 313–317, 315f, 320–322,
    328–329
black holes, 60, 74, 98, 190–191, 250, 295–297
Bohr, Niels, 234
Bose, Satyendra, 215

bosonic fields, 215–218, 219t, 230, 253t, 259,
 293–305, 299f, 311, 323
bosonic wavicles, 249, 323. *See also* bosons
bosons
 gluons as, 230
 Higgs boson, 4–9, 49, 74, 80–81, 92,
  98, 119, 169, 230, 248, 253t, 255–256,
  276–289, 294–305
 identical, 323–326
 photons as, 230, 323–324
 W boson, 230, 236, 246, 253t, 261,
  283–285, 289, 295–299, 302
 Z boson, 230, 253t, 261, 283–286, 289,
  295–299, 302
Brahe, Tycho, 31, 33
Brief History of Time, A (book), 313
Brout, Robert, 283
Buridan, Jean, 61

Carroll, Lewis, 206
cells, 76, 83–87, 231, 241
Cheshire Cat field, 206–208, 214, 277–278
clocks, 30, 126, 133–134, 200–203, 225
CMB (Cosmic Microwave Background),
 161–162, 315
coasting, 29–41, 60–61, 61f, 73, 110, 131,
 165–167, 251–252, 258–259, 275, 318
compass, 170f, 174, 180, 214, 273–274, 273f
composite field, 178, 219t, 283–284
composite particles, 178, 317, 329
compression field, 177–178, 178f, 208, 211t
conservation
 of atoms in chemical reactions, 79–81
 of electric charge, 287
 of energy, 110, 117, 286–287, 301
 of mass in chemical reactions, 79–81
 of quark number, 93–94, 287
Copernicus, Nicolaus, 31, 33–34, 59–60
cosmic certainty limit, 237–239, 288, 295,
 300–303, 311–312, 318, 325, 328. *See also*
  Planck's constant; quantum uncertainty
cosmic fields, 177–179, 183–210, 212–219,
 219t, 233–240, 255–307, 309
cosmic future, 321–322
cosmic liquid, 315–316
cosmic mass-density limit, 295–301
cosmic past, 262, 313–317. *See also* Big Bang

cosmic phase transitions, 315–316, 321
cosmic ray, 43–44
cosmic speed limit, 2, 72–76, 91–93, 99,
 103, 112, 118–120, 156, 198–218, 224,
 236–237, 252. *See also* speed of light
cosmological constant problem, 303, 312
cosmos
 aspects of, 1–17, 74–75, 223–224, 233, 307,
  309, 311–332
 Big Bang and, 86, 313–317, 315f, 320–322,
  328–329
 birth of, 86, 313–317, 315f, 320–322, 328–329
 changeability of, 33–34, 70, 315–317,
  320–322
 electrons and, 2, 92, 120, 316–327, 332
 elements of, 1–10, 223–224, 233, 307
 everywhere-medium and, 165, 178,
  194–200, 211t, 214, 217–218, 219t, 230,
  236, 259, 275
 fields of, 177–179, 183–210, 212–219, 219t,
  233–240, 255–307, 309, 311–329
 future of, 216–218, 282–285, 307–309,
  320–322
 gluons and, 311–316, 319, 327–328
 harmonies of, 130, 281, 290–293
 Higgs field and, 1–7, 39–40, 119, 122,
  138–139, 214, 255–256, 262–263,
  277–278, 297, 301–306, 311–315,
  321–322, 328
 medium of, 194, 251–252
 as musical instrument, 124, 130–131,
  138–139, 247, 262, 272, 289–293
 neutrons and, 309, 311–319, 327–328
 photons and, 161–162, 165–167, 320–324,
  327–328, 332
 protons and, 309, 311–319, 323, 327–328
 quantum field theory and, 309, 311–328
 quarks and, 2, 92, 311–322, 327–328, 332
 relativity principle and, 6–9, 16–19, 30–40,
  70, 161–162, 167, 251, 275–277, 327–329
 resonance and, 125–126, 130, 138–139, 147,
  225–227, 244–250, 256–265, 272–292,
  301
 view of, 1–10, 13–17, 30–33, 40, 50, 74–75,
  191 307
 wavicles and, 309, 311–324, 327–328, 332
 see also universe

Curie, Marie, 100
Curie, Pierre, 100

dark energy, 49, 302
dark matter, 49, 74
decay, 285–289, 303, 316
deep space, 38, 51f, 54–55, 55f, 138, 161–162, 214, 216, 264, 267, 297, 300, 304
Descartes, René, 18
Dialogue Concerning the Two Chief World Systems (book), 18
dissipation
    decay and, 285–289, 316
    explanation of, 109–110
    friction and, 109–110, 131–134
    identity and, 317–319, 318f, 319f
    of motion energy, 109–110, 131–134, 134f
    photons and, 316–319, 319f
    protons and, 316–317, 319
    quantum physics and, 316–319, 318f, 319f
    resonant frequency and, 136–137, 137f
    of vibrations, 131–134, 268, 286–288, 317, 318f
    of waves, 148–149, 149f, 214–215, 232, 286, 288
diversity, of Higgs field's interactions, 280–281, 292–294
diversity, of scientists, 76–78
Doctor Who (TV show), 24–25
down quark, 91–94, 92f, 118, 230, 253t, 261, 283, 285, 287, 319
drag, 35, 161, 166, 196–200, 197f, 203–206, 217, 251
drag method, 196–199, 197f, 206
Dylan, Bob, 128
Dyson, Freeman, 324–325

Earth
    coasting and, 29–41
    Moon and, 12–13, 36, 53–59, 138, 171, 259, 300, 318
    motion and, 1–3, 11–13, 13f, 15–43, 20f, 61f, 327
    orbit of, 12–13, 31–32, 38, 56–58, 57f, 70, 204, 327
    rotation of, 15–22, 30–32, 56–58, 70, 204, 327

speed of, 1–2, 12–13, 13f, 17–25, 20f, 193, 201–202
steady motion and, 1–3, 11–13, 15–43, 20f, 27f, 70
Sun and, 1, 12–13, 25, 29–33, 36–40, 53–60, 70–72, 166–167, 204
earthquake waves, 2–3, 141–142, 149, 165, 182–183, 187, 206, 210, 330–331
Einstein, Albert
    atoms and, 85, 241
    birth and death of, 7, 329
    gravity and, 1, 74, 77, 190, 296, 303–305
    haiku, 251–252, 332
    luminiferous aether and, 162–163, 199–202, 205–210
    philosophical views, 75–76, 292
    predictions and, 222–223
    quantum formula and, 99, 236, 240–248, 251–252
    relativity formula and, 7, 59–60, 81, 99–104, 110–119, 112t, 206, 221–223, 236–248, 247f, 277, 327–328
    relativity principle and, 7–9, 17–18, 37, 70–73, 81, 91–92, 99–104, 110–119, 112t, 189–191, 199–210, 221–223, 234–248, 275–277, 327–329
    rest mass and, 63–78, 97–115, 240–248, 251–252
electric charge, 248, 287
electric field, 130, 155, 169–170, 170f, 176–183, 189, 195, 230, 248, 259–260, 260f, 273–275, 274f, 287. *See also* electromagnetic field
electromagnetic field, 189–209, 211t, 216, 219t, 223–235, 233f, 240–242, 253t, 258, 268, 280–282, 288, 293–294, 318. *See also* electric field; magnetic field
electromagnetic spectrum, 156–157, 157f, 228
electromagnetic waves, 155–158, 157f, 189–209, 211t, 227–228, 234–235, 240–242, 331. *See also* light waves
electron field, 215, 219t, 230–248, 253t, 256–261, 268–288, 290–306, 324
electrons
    atoms and, 5, 80–81, 85f, 87–88, 88f, 90f, 92–95, 97–99, 104–105, 113, 116–121, 121f, 122–124, 166–167, 215–216, 221–222,

electrons *(continued)*
234–241, 248–250, 277, 287–288,
295–298, 306, 316–320, 324–329
cosmos and, 2, 92, 120, 316–327, 332
as elementary particles, 2–8, 95–98, 115,
126, 130, 179, 214, 221–250, 277, 298
Higgs field and, 2, 5–8, 74, 120–122, 121f,
256, 260–261, 265–278, 281–285, 288,
295
identical nature of, 105, 113, 122, 236,
248–249, 317–319, 324–326
mass and, 2, 5–8, 48, 71–76, 81–88, 85f,
88f, 90–98, 90f, 104–105, 110–113,
116–122, 121f
pairs of, 317–319, 319f
quantum physics and, 221–231, 234,
238–239, 246, 311–312
rest mass and, 71–76, 81–88, 85f, 88f,
90–98, 90f, 104–105, 110–113, 116–122,
121f, 235–247, 247f, 248–250, 319–320
stationary electrons, 242–244, 243f,
246–248, 265, 272as wavicles, 230,
235–250, 243f
Elegant Universe, The (book), 76
elementary fields, 178–210, 211t, 212–219,
219t, 230–252, 257–258, 263–283,
291–301, 322, 327–328
elementary particles, 2–8, 95–98, 115, 126–130,
179, 214, 221–252, 238f, 277, 298
elementary waves, 211t, 213–215
empty space
atoms and, 88–90, 166, 192–194, 239–240,
323–325
deep space and, 38, 161–162, 214–216,
300–303
elementary fields and, 195, 208–210, 250–252
expansion of, 60, 191–192, 269, 300–303,
328–329
explanation of, 1–3, 38–39
gravitational field and, 190–191, 201,
206–208, 277–278
gravitational waves and, 190–192, 199–210,
211t
luminiferous aether and, 162–163, 166, 189,
194–196, 199–201, 206–210, 211t
moving through, 1–3, 25, 38, 60, 70–73, 124,
149, 161–167, 194–210, 231, 239–240, 307

nature of, 1–8, 49, 88, 124, 161–167, 190–210,
218, 250–252, 269, 300–303, 328–329
relativity principle and, 192–210, 251–252
vacuum and, 39, 161–162, 300–303
warping of, 1–2, 190–192, 259, 277
energy
calories, 107, 115–116
conservation of, 110, 117, 286–287, 301
dark energy, 49, 302
energy-of-being, 244–247, 247f, 262, 277
explanation of, 106–110
flow of, 131–133, 132f
internal energy, 110–113, 112t, 115–122, 242,
245, 313, 328
invisible forms of, 106–109, 107t, 113
mass and, 49–50, 74, 101–117
matter and, 49–50, 74, 101–102
motion energy, 106–111, 107t, 112t, 116–120,
131–134, 132f, 148–150, 149f, 231–232,
232f, 245, 316, 328
physics energy, 107–110, 235
quantity of, 113, 328
stored energy, 106–113, 107t, 115–122, 242,
245, 313, 328
total energy, 110–114, 112t, 301
vacuum energy, 300–303
visible forms of, 106–107, 107t
energy-of-being, 244–247, 247f, 262, 277
Englert, François, 283
equilibrium, 264–266, 264f, 267f, 271–275,
274f, 301
everywhere-medium, 165, 178, 194–200, 211t,
214, 217–218, 219t, 230, 236, 259, 275

Faraday, Michael, 189
feedback, 298–305, 299f, 312–313
Fermi, Enrico, 215, 324
fermionic field, 215–218, 219t, 230, 249, 275,
292–294, 298–299, 299f, 305, 323–329
fermionic math, 324–326, 329
fermionic wavicles, 249–250, 323–324, 328
fermions, identical, 249, 323–326, 328
field-centric perspective, 185–189, 212, 215,
218, 230–232, 232f, 243, 259, 286
fields
bending field, 174–178, 178f, 183, 208, 210,
211t

bosonic fields, 215–218, 219t, 230, 253t,
259, 293–305, 299f, 311, 323
Cheshire Cat field, 206–208, 214, 277–278
composite fields, 178, 219t, 283–284
compression field, 177–178, 178f, 208, 211t
cosmic fields, 177–179, 183–210, 212–219,
219t, 233–240, 255–307, 309, 311–329
electric field, 130, 155, 169–170, 170f,
176–183, 189, 195, 230, 248, 259–260,
260f, 273–275, 274f, 287
electromagnetic field, 189–209, 211t, 216,
219t, 223–235, 233f, 240–242, 253t,
258, 268, 280–282, 288, 293–294, 318
electron field, 215, 219t, 230–248, 253t,
256–261, 268–288, 290–306, 324
elementary fields, 178–210, 211t, 212–219,
219t, 230–252, 257–258, 263–283,
291–301, 322, 327–328
explanation of, 165–219
fermionic fields, 215–218, 219t, 230, 249, 275,
292–294, 298–299, 299f, 305, 323–329
floppy fields, 262–268, 274f, 275–276, 281,
298, 299f, 302
gluon field, 215–216, 219t, 230–232, 253t,
293–294, 303, 311–316, 319, 327–328
gravitational field, 169, 180, 189–210, 211t,
214–216, 219t, 251–252, 253t, 257–259,
264f, 265–278, 293, 300
Higgs field, 2, 5–8, 39–49, 65, 73–77, 94,
105, 118–122, 130–139, 145, 169, 179–180,
203, 211t, 213–218, 219t, 230, 253t,
255–307, 260f, 264f, 267f, 299f, 311–315,
321–322, 328
interactions of, 177, 212, 214, 259, 274–275,
280, 282–294, 311–312
leaning field, 177–178, 178f, 208, 211t
magnetic field, 130, 155, 169–170, 170f, 174,
179–183, 189, 214–216, 273–274, 273f,
280, 300
magnetization field, 174–176, 175f, 187–188,
187f, 191, 208, 211t, 213, 217, 257, 273,
273f, 277
muon field, 215, 253t, 285, 287, 291, 293
neutrino fields, 215–216, 219t, 230, 253t,
282–287, 293
nonpointing fields, 257–260, 258f, 260f,
275, 283

ordinary fields, 169–183, 213–219, 219t,
226–240, 275–277
orientation field, 273–275
pointing fields, 257–264, 260f, 273–275,
273f, 274f
pressure field, 173–177, 184, 211t, 275,
280radion field, 208, 211t
quark fields, 215–218, 219t, 230–231, 253t,
268, 280–289, 293, 299–302, 311, 322
stiff fields, 262–264, 268, 276–282,
287–288, 298–299, 299f, 301–302, 328
tau field, 215, 219t, 253t, 285, 289, 291, 293
top quark field, 253t, 281, 284, 288, 293,
299, 302, 322
W field, 215, 219t, 230, 236, 253t, 261, 268,
281, 283–285, 289, 293–295, 299, 302
wind field, 171–172, 172f, 173–185, 185f,
186–189, 211t, 212–217, 233, 257, 280, 286
Z field, 215, 219t, 230, 253t, 261, 268, 281,
283–286, 289, 293–295, 299, 302
FitzGerald, George Francis, 101, 205
floppy fields, 262–268, 274f, 275–276, 281,
298, 299f, 302
force
gravitational force, 25, 31–38, 51–62, 61f,
63–66, 74–77, 89–91, 257–261, 258f,
260f, 294–297
Higgs force, 260–261, 288, 294, 311
restoring force, 264–265, 273
sense of touch and, 324–326
strong nuclear force, 91, 92f, 118–120, 294,
311–316
wavicles and, 259–261, 260f
weak nuclear force, 268, 284, 294, 307, 311
frequency
amplitude and, 126–130, 137f, 141–143,
143f, 186, 223–231, 240–241, 243
frequencies of invisible light, 156–157, 157f
frequencies of visible light, 154–161, 157f
frequency ratio, 253t
musical instruments and, 127–130,
135–138, 141, 144–145, 145f, 225–227,
240, 246–247, 256, 262, 269–272
of pendulum, 133–139, 134f, 263–273
resonant frequency, 129–139, 137f, 146–150,
225–227, 245–250, 256–265, 270f,
272–292, 301

frequency *(continued)*
  rest mass and, 126, 244–250, 262–263,
    276, 281, 301, 328
  of vibrations, 126–137, 134f, 135f, 137f,
    138–147, 143f, 225–231, 240–281,
    289–292
friction, 24, 35–38, 44, 52, 60–62, 109–110,
    125, 131–134, 166

galaxies, 12–13, 29, 32–38, 72, 82–86, 117,
    162, 190–204, 297, 309–320
Galilei, Galileo
  birth and death of, 7
  pendulum and, 133
  physics and, 7–8, 40
  planets and, 59, 77
  relativity principle and, 6–7, 17–19, 30–37,
    40, 60–61
Galileo's principle. *See* relativity principle
gamma rays, 156, 157f, 228
Geiger, Hans, 89
general relativity, 190. *See also* gravity; theory
    of gravity
gluon field, 215–216, 219t, 230–232, 253t,
    293–294, 303, 311–316, 319, 327–328
gluons
  cosmos and, 311–316, 319, 327–328
  protons and, 91–95, 92f, 118–119, 215–216,
    236, 248, 280–287
  quarks and, 91–95, 92f, 118–119, 215–216,
    230–232, 248, 280–287, 293, 311–313,
    316, 319, 327–328
"God Particle," 4–5, 255–256. *See also* Higgs
    boson
*God Particle: If the Universe Is the Answer,
    What Is the Question?* (book), 255
GPS (Global Positioning System), 12, 60, 203
gravitational field, 169, 180, 189–210, 211t,
    214–216, 219t, 251–252, 253t, 257–259,
    264f, 265–278, 293, 300
gravitational force, 25, 31–38, 51–62, 51f, 55f,
    57f, 61f, 63–66, 74–77, 89–91, 97, 138,
    193, 257–261, 258f, 260f, 294–297
gravitational mass, 65–66, 73–74, 104–105, 296
gravitational waves, 1–2, 72, 189–193,
    199–210, 211t, 214–216, 236, 323–324
gravitons, 236, 253t, 286–287, 324

gravity
  Einstein and, 1, 74, 77, 190, 296, 303–305
  force of, 25, 31–38, 51–62, 51f, 55f, 57f,
    61f, 63–66, 74–77, 89–91, 97, 138, 193,
    257–261, 258f, 260f, 294–297
  Newton and, 51–59, 107, 190
  pendulum and, 138, 263–273, 264f
  restoring effect and, 264–277, 264f, 267f,
    274f
  weight and, 45, 51–62, 51f, 55f, 57f, 70–71,
    97–102, 138, 150, 241–242, 296–297
Greene, Brian, 76
Grimaldi, Francesco, 166
Gross, David, 311–312
guitar, 126–140, 132f, 135f, 142, 145–148,
    151–152, 157, 162, 214, 225–227, 233,
    240–247, 262, 270–271, 281, 286,
    290–291, 326
Guralnik, Gerald, 283

Hagen, Carl, 283
haiku, 251–252, 332
Halley, Edmond, 59
harmonics, 136, 145–146, 145f
harmonies of the cosmos, 130, 281, 290–293
Hawking, Stephen, 313
Heisenberg's uncertainty principle, 239, 300
Herschel, William, 155
hierarchy puzzle, 297–299, 303–306, 312
Higgs, Peter, 5, 255, 283
Higgs boson, 4–9, 49, 74, 80–81, 92, 98, 119,
    169, 230, 248, 253t, 255–256, 276–289,
    294–305
Higgs field
  in action, 260f, 262–278
  average value of, 213–218, 219t, 256–266,
    273–281, 288, 292, 295–305, 299f,
    315–316, 321–322
  cosmos and, 1–7, 39–40, 119, 122, 138–139,
    214, 255–256, 262–263, 277–278, 297,
    301–306, 311–315, 321–322, 328
  electrons and, 2, 5–8, 74, 120–122, 121f, 256,
    260–261, 265–278, 281–285, 288, 295
  electron field and, 256, 260–261, 265–278,
    281–285, 288, 295, 300–302
  elementary fields and, 179–180, 203,
    213–217, 219t, 262–278, 280, 322

explanation of, 2, 5–8, 39–41, 169–170,
179–180, 255–307
feedback on, 298–305, 299f, 312–313
Higgs boson and, 4–9, 49, 74, 80–81, 92,
98, 119, 169, 230, 248, 253t, 255–256,
276–289, 294–305
Higgs force and, 260–261, 288, 294, 311
Higgs phib and, 6–7, 39–41, 73, 203–204,
263, 276–278
Higgsiferous aether and, 203, 211t, 277
interactions of, 260–261, 275, 280–294,
299–306, 322
neutrons and, 119, 280–284, 296, 303
photons and, 74, 114, 256, 277, 281–288, 303
protons and, 118–121, 280–284, 289,
295–297, 303–306
quarks and, 118–119, 121f, 255, 268, 277,
280–287, 293–299, 302
questions remaining on, 279–307
relativity principle and, 7, 39–41, 256–259,
276–278
resonance and, 130, 138–139, 256,
262–277, 290–293
rest mass and, 65, 73–77, 94, 105, 118–122,
255–256, 259–269, 272–307, 328
as stiffening agent, 262–264, 264f,
265–284, 267f, 298, 322, 328
top quark and, 261, 277, 281–284, 288,
295–299, 302, 322
top quark field and, 281, 284, 288, 299,
302, 322
waves of, 145, 203–204, 211t, 216, 230
wavicles and, 230, 248, 253t, 255–256,
259–269, 276–307
Higgs force, 260–261, 288, 294, 311
Higgs phib, 6–7, 39–41, 73, 203–204, 263,
276–278
Higgs waves, 145, 203–204, 211t, 216, 230
Higgsiferous aether, 203, 211t, 277
Hulse, Russell, 191
human body, 59, 83–89, 115–117, 124, 166,
171, 313, 329
Huygens, Christiaan, 18, 57, 133, 166
hydrogen atoms, 49, 84, 318–319
hydrogen ground state, 318–319, 319f
hypothetical universes, 75. *See also* imaginary
universes

identity, cause of
for protons and neutrons, 248, 317–319
for wavicles, 228, 228f, 234, 236, 248–249,
317–319, 318f, 319f
imaginary universes, 75–76, 209, 211t, 283.
See also hypothetical universes
"impossible sea," 182, 206–207, 210
inertia, 34, 46
infrared light, 154–156, 155f, 157f, 160
interaction tests, 282–289
interactions
of fields, 177, 212, 214, 259, 274–275, 280,
282–294, 311–312
of Higgs field, 260–261, 275, 280–294,
299–306, 322
of ordinary matter, 204, 261, 287–288
of wavicles, 232, 240, 248, 259–261,
281–294, 311
interferometer, 191, 205
intermediaries, 169–171, 195, 216–218, 259,
268, 293–294, 311
internal energy, 110–113, 112t, 115–122, 242,
245, 313, 328
intransigence, 44–54, 46f, 64–70, 67f, 69f,
112–114, 112t, 241–242, 277, 318
intrinsic properties, 53, 63–72, 92, 103–113,
112t, 127, 134, 204–205, 244, 248, 327
intuition, 12, 16, 23–25, 33–35, 52, 62, 81–82,
123–128, 150, 210, 216, 239, 250, 327
inverse square law, 57–58, 259, 311–312, 316
invisible forms of energy, 107–109, 107t, 113
invisible light, 71–73, 155, 160
invisible light, frequencies of, 156–157, 157f
iron, 84, 174–176, 175f, 180–184, 187–193, 187f,
208–210, 211t, 213, 217, 236, 277, 315
isolated bubble, 15–16, 18–19, 25, 40, 70, 161,
196, 199–200, 203–205, 210, 217, 259

Kadanoff, Leo, 77
KAGRA (Kamioka Gravitational Wave
Detector), 191
Kaluza, Theodor, 207–209, 252
Kaluza-Klein modes, 207–209, 211t, 252
Kaufmann, Walter, 100–101
Kepler, Johannes, 25, 31–32, 57, 59–61, 77,
292
Kibble, Thomas, 283

Klein, Oskar, 207–209, 252
known universe, 5, 33, 86, 116. *See also* visible
    universe

laser, 69, 191, 216, 221–223, 227–228, 228f,
    237, 242, 324
Lavoisier, Antoine, 79–80
Leacock, Stephen, 23
leaning field, 177–178, 178f, 208, 211t
Lederman, Leon, 255
LEDs, 115, 221
Lenard, Andrew, 324
LHC (Large Hadron Collider), 4–5, 26, 66–67,
    80, 95, 269, 276, 282–290, 294, 303–307
Lieb, Elliott, 324
light waves, 2, 8, 71–73, 85–86, 85f, 141,
    151–167, 154f, 155f, 157f, 188–189,
    193–206, 215, 223–229, 228f, 235–241,
    327–328. *See also* electromagnetic waves
lightbulbs, 63, 67f, 115–116, 221, 223,
    227–228, 228f, 233f
LIGO (Laser Interferometer Gravitational-
    wave Observatory), 191, 205, 324
liquid crystals, 273–275, 274f
Lorentz, Hendrik, 101, 200, 205–209
luminiferous aether, 162–163, 166, 189,
    194–196, 199–202, 204–210, 211t

magnetic field, 130, 155, 169–170, 170f, 174,
    179–183, 189, 214–216, 273–274, 273f,
    280, 300. *See also* electromagnetic field
magnetization field, 174–176, 175f, 187–188, 187f,
    191, 208, 211t, 213, 217, 257, 273, 273f, 277
Marsden, Ernest, 89
mass
    electrons and, 2, 5–8, 48, 71–76, 81–88,
        85f, 88f, 90–98, 90f, 104–105, 110–113,
        116–122, 121f
    energy and, 49–50, 74, 101–117
    explanation of, 2, 43–120, 46f, 51f, 55f
    gravitational mass, 65–66, 73–74, 104–105,
        296
    intransigence and, 44–54, 46f, 64–70, 67f,
        69f, 112–114, 112t, 241–242, 277, 318
    invariant mass, 65
    mass density, 174–177, 208, 211t, 257,
        273–274, 295–301, 329

mass hierarchy, 297–299, 303–306, 312
mass ratio, 253t
matter and, 47–50, 64, 79, 98, 113, 328–329
neutrons and, 74–76, 81, 97–98, 108,
    113–120, 313–319
Newton and, 50–59, 98–101, 107, 113, 241,
    328–329
photons and, 71–74, 89–92, 98, 108,
    116–118, 235–236
Planck mass, 277, 295–299, 302
protons and, 66, 74–76, 81, 92f, 97–98, 104,
    108, 113–120, 121f, 313–319
quarks and, 115–122, 121f
relativistic mass, 65–66, 67f, 69f, 103–105,
    112t, 114
rest mass, 63–78, 67f, 69f, 94–113, 112t,
    115–122, 121f, 126, 235–250, 247f, 253t,
    255–269, 276–307, 313–322, 328
wavicles and, 234–250, 253t
weight and, 51–62, 51f, 55f, 57f, 70, 97
see also rest mass
mass density, 174–177, 208, 211t, 257,
    273–274, 295–301, 329
mass hierarchy, 297–299, 303–306, 312
material, structure of, 79–96
Mathematical Principles of Natural Philosophy
    (book), 59
matter
    antimatter, 92–94, 316
    atoms and, 49–50, 79, 94–98, 113, 328–329
    dark matter, 49, 74
    energy and, 49–50, 74, 101–102
    mass and, 47–50, 64, 79, 98, 113, 328–329
    Newton and, 50, 98–101, 113, 328–329
    ordinary matter, 50, 204, 261, 287–288
    quantity of, 50, 64, 79, 98, 113, 328–329
Maxwell, James Clerk, 189
medium, 148–150, 162–163, 165–167, 173–178,
    175f, 177f, 182–210, 185f, 187f, 197f,
    211t, 212–214, 217–218, 219t, 230–232,
    236, 243, 251–252, 259, 269, 270f, 271f,
    273–275, 274f, 277, 314
medium-centric perspective, 184–188, 232,
    259
Michelson, Albert, 205
microwaves, 71, 111, 156, 157f, 161–162, 239,
    315, 331

Milky Way (galaxy), 12–13, 29, 32, 72, 86, 193, 201–204, 314

Moon
Earth and, 12–13, 36, 53–59, 138, 171, 259, 300, 318
gravity of, 54–59, 138
Newton and, 55–60, 166
orbit of, 55–56, 58–59, 166
speed of, 56–57
speed of light and, 71–72
weight on, 54–55

Morley, Edward, 205

motion
ambimotional object, 23, 29, 200
Earth and, 1–3, 11–13, 13f, 15–43, 20f, 61f, 327
energy and, 106–111, 107t, 112t, 116–120, 131–134, 132f, 134f, 148–150, 149f, 231–232, 232f, 245, 316, 328
laws of, 34–36
Newton and, 34–36, 51–61, 61f, 107
omnimotional object, 23, 29, 200
polymotional object, 23, 29, 67, 200
steady motion, 1–3, 11–13, 15–43, 20f, 27f, 63–71, 131–132, 193, 205

motion energy, 106–111, 107t, 112t, 116–120, 131–134, 132f, 134f, 148–150, 149f, 231–232, 232f, 245, 316, 328

MSM (Minimal Standard Model), 268–269, 280–285, 290, 322

muon field, 215, 253t, 285, 287, 291, 293

muons, 43–44, 215, 219t, 253t, 285–287, 291–293

musical chord, 144, 157, 290–291

musical instruments
chords and, 144, 157, 290–291
cosmos as musical instrument, 124, 130–131, 138–139, 247, 262, 272, 289–293
frequency and, 127–130, 135–138, 141, 144–145, 145f, 225–227, 240, 246–247, 256, 262, 269–272
harmonics and, 136, 145–146, 145f
pendulum and, 133–139
resonance and, 125–150, 225–227, 240, 246–247, 256, 262, 269–272
sound waves and, 214–215, 233

vibrations and, 125–139, 132f, 135f, 162, 225–227, 240–248, 269–272, 270f, 271f, 286–291
see also guitar; piano; string instruments

naturalness puzzle, 299

neutrino fields, 215–216, 219t, 230, 253t, 282–287, 293

neutrinos, 49, 240, 246, 250, 256, 323

neutron star, 89–90, 191, 240

neutrons
atomic nucleus and, 87–90, 88f, 90f, 97–98
cosmos and, 309, 311–319, 327–328
elementary fields and, 215–216
Higgs field and, 119, 280–284, 296, 303
mass and, 74–76, 81, 97–98, 108, 113–120, 313–319
origin of, 313–317
quantum physics and, 239–240, 248
quarks and, 94–95, 215–216, 248, 309, 311–316, 327–328

Newton, Isaac
birth and death of, 7
gravity and, 51–59, 107, 190
light waves and, 165–166, 203
mass and, 50–59, 98–101, 107, 113, 241, 328–329
matter and, 50, 98–101, 113, 328–329
motion and, 34–36, 51–61, 61f, 107
physics and, 7, 40, 75–80, 251, 292, 328–329
predictions and, 222–223
relativity principle and, 18, 34–36, 60–62, 64–66, 70–71
weight and, 51–59, 70

nightmare property, 198–208, 210, 252

Nobel Prize, 89, 100, 191, 241, 255, 283, 312

nonpointing fields, 257–260, 258f, 260f, 275, 283

nonzero rest mass, 73–74, 235, 250, 253t, 268, 284

nuclear explosions, 100, 116–117, 121, 190

nucleus. *See* atomic nucleus

ocean waves, 1–2, 148, 165, 173, 176, 187, 223, 238, 330

omnimotional object, 23, 29, 200

orbitals, 249–250, 324–325
orbits, 12–13, 31–32, 38, 55–63, 57f, 61f, 70,
    77, 166, 191, 251, 304, 327
ordinary fields, 169–183, 213–219, 219t,
    226–240, 275–277
ordinary material, 1–3, 49, 81, 83–84, 90, 120,
    194, 200, 204–205, 231–232, 261, 285,
    287–288, 325
ordinary matter, 50, 204, 261, 287–288
ordinary medium, 173–174, 182, 190–207,
    197f, 214, 217–219, 219t, 236, 269–275,
    314
ordinary waves, 198, 211t
orientation field, 273–275
outer space, 12, 24, 36–38, 43, 60, 120,
    161–162, 165–166, 331
oxygen atoms, 248–249, 317

particle accelerators, 4–5, 66–67, 120, 246,
    289, 303–307. See also LHC
particles
    antiparticles, 91–94, 230–231
    composite particles, 178, 284, 317, 328
    decay of, 285–289, 303, 316
    electrons as, 2–8, 95–98, 115, 126, 130, 179,
        214, 221–250
    elementary particles, 2–5, 95–98, 115,
        126–130, 179, 214, 229–238, 238f,
        246–252, 277, 298
    "God Particle," 4–5, 255–256
    particle accelerators, 4–5, 66–67, 120, 246,
        303–307
    quantum physics and, 221–253
    subatomic particles, 16, 23, 39–40, 82, 108,
        117–120, 250, 261, 320–322, 329
    trapping, 118–120, 215, 312, 316
    waves and, 123, 229–231, 252
    versus wavicles, 229–231, 235
particulate wave, 228f, 229–231, 252,
    323–325
Pauli Exclusion Principle, 249, 323–325
pendulum
    clocks and, 133–134
    frequency of, 133–139, 134f, 263–273
    gravity and, 138, 263–273, 264f
    musical instruments and, 133–139
Penzias, Arno, 161

phase transitions, 315–316, 321
phibs, 6–7, 9, 39–41, 73, 91–94, 119, 203–204,
    263, 276–278, 313–314
Philosophiæ Naturalis Principia Mathematica
    (book), 59
photo-electric effect, 241
photons
    CMB, 161–162, 315
    cosmos and, 161–162, 165–167, 320–324,
        327–328, 332
    dissipation and, 316–319, 319f
    Higgs field and, 74, 114, 256, 277, 281–288,
        303
    intrinsic properties of, 92, 244, 248
    mass and, 71–74, 89–92, 98, 108, 116–118,
        235–236
    quantum formula and, 237, 240–241
    quantum physics and, 221–248, 228f
    sense of sight and, 71, 156
    waves and, 156, 161–165, 216, 221–228,
        228f, 253t
piano, 127–130, 136–138, 141, 156–157
Planck, Max, 99–100, 237, 295
Planck length, 295–296
Planck mass, 277, 295–299, 302
Planck's constant, 99–100, 236–240,
    245–246, 300. See also cosmic certainty
    limit
Poincaré, Henri, 101, 205
pointing fields, 257–264, 260f, 273–275,
    273f, 274f
Politzer, David, 311–312
polymotional object, 23, 29, 67, 200
positrons, 92, 94, 98, 231, 253t, 288
pressure field, 173–177, 184, 211t, 275, 280
pressure waves, 206, 211t
principle of relativity. See relativity principle
protons
    atomic nucleus and, 87–90, 88f, 90f, 97–98
    cosmos and, 309, 311–319, 323, 327–328
    dissipation and, 316–317, 319
    elementary fields and, 178, 215–216
    gluons and, 91–95, 92f, 118–119, 215–216,
        236, 248, 280–287
    Higgs field and, 118–121, 280–284, 289,
        295–297, 303–306
    identical nature of, 97, 309, 317–320

mass and, 66, 74–76, 81, 92f, 97–98, 104, 108, 113–120, 121f, 313–319
origin of, 313–317
quantum physics and, 230, 238–242, 248, 311–312
quarks and, 91–95, 92f, 118–120, 121f, 215–216, 230, 248, 280–287, 289, 311–319, 322–323, 327–328
wavicles and, 237–239, 238f, 311–312, 318–319

quanta (*plural of* quantum), 226–248
quantum field theory
atoms and, 320–326
cosmos and, 309, 311–328
explanation of, 131, 221–253
vacuum energy and, 300–303
wavicles and, 221–223, 234–235, 240, 303, 311–312, 326
quantum formula, 99, 236–248, 247f, 251–252
quantum physics
cosmic future and, 321–322
in daily life, 43, 75–80, 226–227, 300–307, 309, 322–329
dissipation and, 316–319, 318f, 319f
electrons and, 221–231, 234, 238–239, 246, 311–312
explanation of, 221–253, 276–277, 309, 311
hierarchy puzzle and, 297–299, 303–306, 312
neutrons and, 239–240, 248
particles and, 221–253
photons and, 221–248, 228f
protons and, 230, 238–242, 248, 311–320
quarks and, 221–229, 246–250
relativity principle and, 276–277, 327–329
sense of touch and, 322–326
uncertainty and, 239, 288, 300, 311, 321–325, 328–329
wavicles and, 233–253
quantum uncertainty, 239, 288, 300, 311, 321–325, 328–329. *See also* cosmic certainty limit
quantum uncertainty principle, 239, 300
quark fields, 215–218, 219t, 230–231, 253t, 268, 280–289, 293, 299–302, 311, 322

quarks
anti-quarks and, 91–95, 92f, 118–119, 230–232, 253t, 261, 287–289, 312, 316, 319
cosmos and, 2, 92, 311–322, 327–328, 332
down quark, 91–94, 92f, 118, 230, 253t, 261, 283, 285, 287, 319
explanation of, 2, 91–98, 92f
gluons and, 91–95, 92f, 118–119, 215–216, 230–232, 248, 280–287, 293, 311–313, 316, 319, 327–328
Higgs field and, 118–119, 121f, 255, 268, 277, 280–287, 293–299, 302
mass and, 115–122, 121f
neutrons and, 94–95, 215–216, 248, 309, 311–316, 327–328
protons and, 91–95, 92f, 118–120, 121f, 215–216, 230, 248, 280–287, 289, 295–297, 311–319, 322–323, 327–328
quantum physics and, 221–229, 246–250
rest mass and, 63–78, 67f, 69f, 94–113, 112t, 115–122, 121f, 126, 235–250, 247f, 253t, 261, 281–282, 295–297, 319–322
strange quark, 91–93, 104, 253t, 289
top quark, 250, 253t, 261, 277, 281–284, 287–288, 293, 295–299, 302, 322
up quark, 91–94, 92f, 230–231, 253t, 283, 319

radio waves, 71–73, 156, 157f, 160–162, 179, 228, 239–241
radioactivity, 43, 89, 100, 117
radion field, 208, 211t
rainbows, 153–155, 155f, 235, 331
raindrops, 153–154, 154f, 236
reflection, 152–154, 154f
refraction, 153–154, 154f
relativistic mass, 65–66, 67f, 69f, 103–105, 112t, 114
relativity formula, 7, 59–60, 81, 99–104, 110–119, 112t, 206, 221–223, 236–248, 247f, 277, 327–328
relativity principle
coasting and, 30–37, 167
cosmos and, 6–9, 16–19, 30–40, 70, 161–162, 167, 251, 275–277, 327–329
empty space and, 192–210, 251–252

relativity principle *(continued)*
  Einstein and, 7–9, 17–18, 37, 70–73, 81,
    91–92, 99–104, 110–119, 112t, 189–191,
    199–210, 221–223, 234–248, 275–277,
    327–329
  explanation of, 6–7, 15–28, 43, 60–62,
    190–218
  Galileo and, 6–7, 17–19, 30–37, 40, 60–61
  Higgs field and, 7, 39–41, 256–259,
    276–278
  Newton and, 18, 34–36, 60–62, 64–66,
    70–71
  quantum physics and, 276–277, 327–329
  rest mass and, 69–70
resonance
  cosmos and, 125–126, 130, 138–139, 147,
    225–227, 244–250, 256–265, 272–292,
    301
  explanation of, 125–139
  frequency and, 129–139, 137f, 146–150,
    225–227, 245–250, 256–265, 270f,
    272–292, 301
  Higgs field and, 130, 138–139, 256,
    262–277, 290–293
  musical instruments and, 125–150, 225–227,
    240, 246–247, 256, 262, 269–272
  rest mass and, 244–250, 262–263, 276
resonant frequency, 129–139, 137f, 146–150,
    225–227, 245–250, 256–265, 270f,
    272–292, 301
rest mass
  cosmic past and, 262, 313–317
  decay and, 285–289, 316
  Einstein and, 63–78, 97–115, 240–248,
    251–252
  electrons and, 71–76, 81–88, 85f, 88f,
    90–98, 90f, 104–105, 110–113, 116–122,
    121f, 235–247, 247f, 248–250, 319–320
  energy-of-being and, 244–247, 247f, 262, 277
  explanation of, 63–78
  frequency and, 126, 244–250, 262–263,
    276, 281, 301, 328
  Higgs field and, 65, 73–77, 94, 105, 118–122,
    255–256, 259–269, 276–307, 328
  internal energy and, 110–122, 245, 313, 328
  intransigence and, 64–70, 67f, 69f, 112–114,
    112t, 241–242, 277, 318

  as an intrinsic property, 63–72, 111–113,
    248
  nonzero rest mass, 73–74, 235, 250, 253t,
    268, 284
  Planck mass and, 277, 295–299, 302
  quarks and, 63–78, 67f, 69f, 94–113, 112t,
    115–122, 121f, 126, 235–250, 247f, 250,
    253t, 261, 281–282, 295–297, 319–322
  relativistic mass and, 65–66, 67f, 103–105,
    112t, 114
  relativity principle and, 69–70
  resonance and, 244–250, 262–263, 276
  rule of decreasing rest mass, 287–288
  stored energy and, 110–122, 242, 245, 313,
    328
  wavicles and, 234–250, 243f, 247f, 253t,
    255–269, 276–307, 319–320
  zero rest mass, 71–74, 118–121, 235–236,
    282, 285, 302
  see also mass
resting law, 36–40, 60, 131
restoring effect, 264–277, 264f, 267f, 274f
restoring force, 264–265, 273
Ritter, Johan, 155
rule of decreasing rest mass, 287–288
Rutherford, Ernest, 89–90

Salam, Abdus, 298
Schrödinger wave function, 231
seismic waves. *See* earthquake waves
senses
  abilities of, 171–181
  failings of, 192
  interactions of, 89
  sense of hearing, 3, 127–128, 140, 151–163,
    291
  sense of scale, 81–82
  sense of sight, 3, 71, 151–163
  sense of smell, 44
  sense of touch, 3, 322–326
shells, 249–250, 324–325
simple waves, 140–145, 141f, 143f, 145f, 148,
    215, 225–229, 228f
Sina, Abu ʿAli ibn, 61
sound waves, 2–3, 8, 26, 132–133, 132f,
    140–162, 165, 182–206, 185f, 211t, 213–216,
    229–233, 243–244, 286–292, 330

space
  bending of, 1–2, 74
  contraction of, 320
  expansion of, 32, 49, 60, 191–192, 262,
    269, 300–303, 313–316, 315f, 320,
    328–329
  space resistance, absence of, 1–3, 196
  time and, 1–2, 8, 17, 37, 101, 201–213
  see also empty space
space resistance, absence of, 1–3, 196
speed
  of airplanes, 24–28, 28f, 38f, 204
  of Earth, 1–2, 12–13, 13f, 17–25, 20f, 193,
    201–202
  of gravitational waves, 1–2, 72, 199
  of Higgs waves, 145, 204
  of light, 2, 71–72, 199–203, 236
  of Moon, 56–57
  relativity of, 18–27, 20f, 27f, 28f
  of sound, 26, 155–156, 198–204
  see also cosmic speed limit
spin waves, 183, 187–188, 187f, 191, 210, 211t,
  248
standing waves, 145–147, 145f, 162, 215, 218,
  225–227, 242–246, 243f, 263–271, 270f,
  271f, 272–276, 274f, 289–291, 302
stars, 5, 12, 16, 30–33, 36–39, 44, 49, 54–56,
  64, 72–76, 80, 86, 89–90, 117, 160–162,
  166, 191, 233, 240, 248, 256, 297, 300,
  304, 309, 314, 331
static electricity, 170, 170f
stationary electrons, 242–244, 243f,
  246–248, 265, 272
stiff fields, 262–264, 268, 276–282, 287–288,
  298–299, 299f, 301–302, 328
stiffening agents, 262–264, 264f, 265–284,
  267f, 298, 322, 328
stored energy, 106–113, 107t, 115–122, 242,
  245, 313, 328. *See also* internal energy
strange quark, 91–93, 104, 253t, 289
string instruments
  chords and, 144, 157, 290–291
  frequency and, 127–130, 135–138, 141,
    144–145, 145f, 225–227, 240, 246–247,
    256, 262, 269–272
  guitar, 126–140, 132f, 135f, 142, 145–148,
    151–152, 157, 162, 214, 225–227, 233,

    240–247, 262, 270–271, 281, 286,
    290–291, 326
  harmonics and, 136, 145–146, 145f
  Higgs field and, 138–139, 269–272
  pendulum and, 133–139
  piano and, 127–130, 136–138, 141, 156–157
  resonance and, 125–150, 225–227, 240,
    246–247, 256, 262, 269–272
  sound waves and, 151–152, 162, 214–215,
    233, 286
  vibrations and, 132f, 133–139, 135f, 162,
    225–227, 240–248, 269–272, 270f, 271f,
    286–291
string theory, 130–131, 208, 222
strong nuclear force, 91, 92f, 118–120, 294,
  311–316
Styrenian fable, 45–47, 46f, 50, 328
subatomic particles, 16, 23, 39–40, 82, 108,
  117–120, 250, 261, 320–322, 329
Sun
  coasting law and, 60–61, 61f
  Earth and, 1, 12–13, 25, 29–33, 36–40,
    53–60, 70–72, 166–167, 204
  observations of, 31–32, 57–60
  orbit around, 31–32, 56, 60, 61f, 70, 204
  speed of light and, 71–72
  as star, 32
  temperature of, 316
  waves and, 148, 166–167
swings, 125–134, 146, 317–318, 318f

tau field, 215, 219t, 253t, 285, 289, 291, 293
Taylor, Joseph, 191
Teresi, Dick, 255
theory of gravity, 77, 296. *See also* general
  relativity; gravity
theory of relativity, 221–223. *See also* Einstein,
  Albert
Thirring, Walter, 324
Thomas, Llewellyn, 324
time, and distance, 201, 301
time, and space, 1–2, 8, 17, 37, 101, 201–213
top quark, 250, 253t, 261, 277, 281–284,
  287–288, 293, 295–299, 302, 322
top quark field, 253t, 281, 284, 288, 293, 299,
  302, 322
total energy, 110–114, 112t, 301

trajectories, 234, 237–239, 238f, 300
traveling waves, 142–143, 143f, 144f, 145–149,
　　149f, 162, 186, 215, 218, 225–226, 228f,
　　231–233, 232f, 242, 245, 265, 269–271,
　　270f, 271f, 289–291

ultraviolet light, 73, 154–156, 155f, 157f, 241
universal firestorm, 314–315, 315f
universe
　　aspects of, 1–17, 74–75, 223–224, 233, 307,
　　　　309, 311–332
　　Big Bang and, 86, 313–317, 315f, 320–322,
　　　　328–329
　　birth of, 86, 313–317, 315f, 320–322, 328–329
　　changeability of, 33–34, 70, 315–317,
　　　　320–322
　　elements of, 1–10, 223–224, 233, 307
　　future of, 216–218, 282–285, 307–309,
　　　　320–322
　　Higgs field and, 1–7, 39–40, 119, 122,
　　　　138–139, 214, 255–256, 262–263,
　　　　277–278, 297, 301–306, 311–315,
　　　　321–322, 328
　　hypothetical universes, 75
　　imaginary universes, 75–76, 209, 211t, 283
　　known universe, 5, 33, 86, 116
　　as musical instrument, 124, 130–131,
　　　　138–139, 247, 262, 272, 289–293
　　relativity principle and, 6–9, 16–19, 30–40,
　　　　70, 167, 251, 275–277, 327–329
　　resonance and, 125–126, 130, 138–139, 147,
　　　　225–227, 244–250, 256–265, 272–292,
　　　　301
　　view of, 1–10, 13–17, 30–33, 40, 50, 74–75,
　　　　191, 307,
　　visible universe, 5, 33, 86, 116
　　wavicles and, 309, 311–324, 327–328, 332
　　see also cosmos
up quark, 91–94, 92f, 230–231, 253t, 283, 319

vacuum, 39, 161–162, 300–303
vacuum energy, 300–303
vibrations
　　amplitude and, 126–130, 225–231,
　　　　240–241, 243
　　dissipation of, 131–134, 268, 286–288, 317,
　　　　318f

explanation of, 126–130
frequency of, 126–137, 134f, 135f, 137f,
　　138–147, 143f, 225–231, 240–281,
　　289–292
musical instruments and, 125–139, 132f,
　　135f, 162, 225–227, 240–248, 269–272,
　　270f, 271f, 286–291
springs and, 133, 137f, 266–268, 267f
waves and, 140–141, 225–228
Virgo experiment, 191
visible forms of energy, 106–107, 107t
visible light, 71–73, 85–86, 85f, 153–156, 155f,
　　157f, 158–166, 193, 235, 241
visible light, frequencies of, 154–161, 157f
visible universe, 5, 33, 86, 116. *See also* known
　　universe

W boson, 230, 236, 246, 253t, 261, 283–285,
　　289, 295–299, 302
W field, 215, 219t, 230, 236, 253t, 261, 268,
　　281, 283–285, 289, 293–295, 299, 302
water pressure, 173, 176
wave function, 231
wave speed method, 196–206, 197f
waves
　　crests and troughs of, 140–150, 141f, 143f,
　　　　145f, 184–185, 185f, 198, 223–228, 243,
　　　　269–271
　　dissipation of, 148–149, 149f, 214–215, 232,
　　　　286, 288
　　electromagnetic waves, 155–158, 157f,
　　　　189–209, 211t, 227–228, 234–235,
　　　　240–242, 331
　　elementary waves, 211t, 213–215
　　explanation of, 1–8, 123–165
　　gravitational waves, 1–2, 72, 189–193,
　　　　199–210, 211t, 214–216, 236, 323–324
　　of Higgs field, 145, 203–204, 211t, 216, 230
　　in "impossible sea," 182, 206–207, 210
　　light waves, 2, 8, 71–73, 85–86, 85f,
　　　　141, 151–167, 154f, 155f, 157f, 188–189,
　　　　193–206, 215, 223–229, 228f, 235–241,
　　　　327–328
　　medium of, 148–150, 162–165, 173–178,
　　　　182–207, 197f, 199–210, 211t, 214
　　microwaves, 71, 111, 156, 157f, 161–162, 239,
　　　　315, 331

neutrons and, 123–124
ocean waves, 1–2, 148, 165, 173, 176, 187,
    223, 238, 330
ordinary waves, 198, 211t
particles and, 123, 229–231, 252particulate
    waves, 228f, 229–231, 252, 323–325
photons and, 156, 161–165, 216, 221–228,
    228f, 253t
pressure waves, 206, 211t
radio waves, 71–73, 156, 157f, 160–162, 179,
    228, 239–241
seismic waves, 2–3, 141–142, 149, 165,
    182–183, 187, 206, 210, 330–331
simple waves, 140–145, 141f, 143f, 145f, 148,
    215, 225–229, 228f
sound waves, 2–3, 8, 26, 132–133, 132f,
    140–162, 165, 182–206, 185f, 211t,
    213–216, 229–233, 243–244, 286–292,
    330
spin waves, 183, 187–188, 187f, 191, 210,
    211t, 248
standing waves, 145–147, 145f, 162, 215, 218,
    225–227, 242–246, 243f, 263–271, 270f,
    271f, 272–276, 274f, 289–291, 302
traveling waves, 142–143, 143f, 144f,
    145–149, 149f, 162, 186, 215, 218,
    225–226, 228f, 231–233, 232f, 242, 245,
    265, 269–271, 270f, 271f, 289–291
vibrations and, 140–141, 225–228
wavicles and, 228f, 229–233, 233f,
    323–325
wavicles
bosonic wavicles, 249, 323
cosmos and, 309, 311–324, 327–328, 332
decay of, 285–289, 303
elementary particles and, 229–232, 235
energy-of-being and, 244–247, 247f, 262,
    277
explanation of, 229–232, 242–244, 243f
fermionic wavicles, 249–250, 323–324,
    328
force and, 259–261, 260f
Higgs field and, 230, 248, 253t, 255–256,
    259–269, 276–307
identical, 228, 228f, 234, 236, 248–250,
    252, 323–324, 328
identity and, 317–319, 318f, 319f

interactions of, 232, 240, 248, 259–261,
    281–294, 311
versus particles, 229–231, 235
as particulate waves, 228f, 229–231, 243f,
    252, 323–325
Pauli Exclusion Principle and, 249, 323–325
protons and, 237–239, 238f, 311–312,
    318–319
quantum field theory and, 221–223,
    234–235, 240, 303, 311–312, 326
quantum physics and, 233–253
rest mass and, 234–250, 243f, 247f, 253t,
    255–269, 276–307, 319–320
trajectories of, 234, 237–239, 238f, 300
waves and, 228f, 229–233, 233f, 323–325
weak nuclear force, 268, 284, 294, 307, 311
weight, 45, 51–62, 51f, 55f, 57f, 70–71,
    97–102, 138, 150, 241–242, 296–297. *See
    also* gravity
Weinberg, Steven, 298
Wilczek, Frank, 311–312
Wilson, Kenneth, 77
Wilson, Robert, 161
wind
calm wind, 213–217
map of, 172f
measurement of, 171–189, 213–214
nature of, 1, 27, 46, 46f, 50, 107, 151,
    171–172, 172f, 173–185, 185f, 186–194,
    211t, 212–217, 233, 257, 280, 286, 313
space wind, absence of, 1
wind field, 171–172, 172f, 173–185, 185f,
    186–189, 211t, 212–217, 233, 257, 280, 286
"Wind Map," 172f
wind meter, 171–175, 184

X-rays, 71, 73, 89, 156, 157f, 162, 171, 179,
    240–241, 323

Young, Thomas, 166

Z boson, 230, 253t, 261, 283–286, 289,
    295–299, 302
Z field, 215, 219t, 230, 253t, 261, 268, 281,
    283–286, 289, 293–295, 299, 302
zero rest mass, 71–74, 118–121, 235–236, 282,
    285, 302

Matt Strassler is a theoretical physicist, blogger, and writer whose research often takes him to the Large Hadron Collider. An associate of the Harvard University Physics Department and a former member of the Institute for Advanced Study, he was previously a professor at the University of Pennsylvania, the University of Washington, and Rutgers University. He lives in rural Massachusetts.